Dynamical Systems - Latest Developments and Applications

Edited by Mohammad Shamsuzzoha and G. Lloyds Raja

Published in London, United Kingdom

Dynamical Systems - Latest Developments and Applications
http://dx.doi.org/10.5772/intechopen.1003418
Edited by Mohammad Shamsuzzoha and G. Lloyds Raja

Contributors
Alexander V. Chernodarov, Alvaro Humberto Salas, Amir Aliakbari, Ayadi Souad, Boris Atenas, Catalin Gabriel Dumitras, Cristina Maricela Rusu, Edward Larroza, Emanuel Nazaretian, Florin Nedeff, Godwin Sani, Iosif Ioja, Maricel Agop, Peiman Amiri, Uchenna Diala, Valentin Nedeff, Verónica Bahoz, Vlad Ghizdovat, Zeraoulia Rafik

First published in London, United Kingdom, 2025 by IntechOpen
IntechOpen is the global imprint of INTECHOPEN LIMITED, registered in England and Wales, registration number: 11086078, 167-169 Great Portland Street, London, W1W 5PF, United Kingdom

For EU product safety concerns: IN TECH d.o.o., Prolaz Marije Krucifikse Kozulić 3, 51000 Rijeka, Croatia, info@intechopen.com or visit our website at intechopen.com.

British Library Cataloguing-in-Publication Data
A catalogue record for this book is available from the British Library

Dynamical Systems - Latest Developments and Applications
Edited by Mohammad Shamsuzzoha and G. Lloyds Raja
p. cm.
Print ISBN 978-1-83634-368-4
Online ISBN 978-1-83634-367-7
eBook (PDF) ISBN 978-1-83634-369-1

Meet the editors

Dr. Shamsuzzoha has a Ph.D. in chemical engineering specializing in process design, modeling, simulation, and control. He conducts studies and provides process engineering, modeling, and simulation support to clients in a wide range of process industries. Prior to his current role, he held the position of Senior Specialist at ADNOC Refining Co., where he focused on providing technical support to refineries and gas plants. This included process modeling and simulation, as well as the modification and design of refinery units and associated gas plants. Recently, he has completed an MBA in Entrepreneurship & Leadership, and he has unique work experience in academics, industries, and research & development in a multicultural environment. He is a pioneering researcher and practicing engineer in process modelling, simulation and control.

Dr. G. Lloyds Raja was born in Kanyakumari, India. He received his bachelor's and master's degrees in Engineering from Anna University. He was awarded a Ph.D. from the Indian Institute of Technology Patna for his dissertation entitled "Enhanced Cascade Control Strategies for Unstable and Industrial Processes". For a brief period, he was a postdoctoral researcher at the Department of Automation in Shanghai Jiao Tong University, China. Presently, he is an Assistant Professor at the National Institute of Technology, Patna. He has over 10 years of experience in teaching and research. During this period, he has authored or co-authored nearly 75 papers in international journals and peer-reviewed conference proceedings. He has four granted patents and guided many doctoral and post-graduate dissertations. He has taught courses such as linear control systems, modern control theory, electronic circuits, and industrial process control, and delivered invited lectures in several forums.

Contents

Preface

The study of dynamical systems continues to evolve as new theories, methodologies, and applications emerge across science and engineering disciplines. With the growing complexity of modern systems—ranging from nonlinear control mechanisms and energy dissipation in mechanical systems to agent-based modeling and multifractal analysis—there is a clear need to revisit fundamental principles while embracing cutting-edge perspectives. This book, *Dynamical Systems – Latest Developments and Applications*, aims to bridge this evolving landscape by presenting a collection of chapters that reflect both theoretical advances and practical insights.

Chapter 1, "Perspective Chapter: Optimization of Observations Based on Non-Quadratic Criteria and Controlled Error Compensation in Nonlinear Dynamical Systems", opens the volume with a foundational perspective. It explores the limitations of traditional quadratic criteria and introduces novel optimization frameworks for enhancing observation accuracy in nonlinear environments, laying a rigorous groundwork for advanced state estimation and control techniques.

Chapter 2, "Parametric Study of the Energy Dissipation by a Suspension System", shifts the focus to mechanical engineering applications. Through a detailed parametric analysis, it investigates how suspension systems can be optimized to maximize energy dissipation and stability, which is vital for automotive and structural vibration control.

Chapter 3, "Structural Stability of Materials", addresses the critical issue of material behavior under various loading and environmental conditions. This chapter integrates concepts from dynamical systems theory with materials science to evaluate and predict stability thresholds, which are essential for the design of durable engineering structures.

Chapter 4, "Holographic-Type Behaviors in Complex Systems Dynamics from a Multifractal Perspective of Motion", introduces a novel lens for understanding complexity. It proposes a multifractal framework to interpret the dynamic evolution of complex systems, suggesting holographic-type representations as a means to uncover hidden symmetries and emergent behaviors.

Chapter 5, "Agent-Based Models", delves into the modeling of systems composed of interacting autonomous agents. From biological swarms to socio-economic simulations, this chapter highlights how decentralized, rule-based interactions can give rise to rich, system-level dynamics.

Finally, Chapter 6, "Chaotic Dynamics Derived from the Montgomery Conjecture: Application to Electrical Systems", examines the chaotic dynamics that arise from the Montgomery Conjecture, bridging the gap between abstract mathematical theory and practical applications in electrical systems. Through rigorous modeling and simulation, the authors uncover intricate patterns and behaviors that deepen our understanding of deterministic chaos.

This compilation is intended for researchers, advanced students, and practitioners seeking a deeper understanding of the modern landscape of dynamical systems. By bringing together diverse yet interconnected topics, this volume aims to inspire interdisciplinary collaboration and encourage novel applications in both theory and practice.

We hope this book serves as a valuable resource and sparks further innovation in the dynamic world of dynamical systems.

Dr. Mohammad Shamsuzzoha
Principle Process Engineer,
Billington Process Technology,
Sandvika, Norway

Dr. G. Lloyds Raja
Assistant Professor,
Department of Electrical Engineering,
National Institute of Technology Patna,
Patna, Bihar

Chapter 1

Perspective Chapter: Optimization of Observations Based on Non-Quadratic Criteria and Controlled Error Compensation in Nonlinear Dynamical Systems

Alexander V. Chernodarov

Abstract

This section is devoted to the problem of joint estimation and control of parameters of nonlinear dynamic systems (NDS) under stochastic conditions. The errors of the NDS with respect to the reference phase path, which is determined by external observations, are considered as such parameters. When including the extended Kalman filter (EKF) in the estimation loop, the control must limit the change in NDS errors to a linear region relative to the reference phase path. The implementation of this approach to damping error estimates is associated with the simultaneous solution of nonlinear differential equations of the NDS and linearized error equations. It is proposed to determine the damping coefficients of error estimates based on the method of inverse dynamics problems. The chapter also discusses the problem of increasing the reliability of estimates of NDS errors in conditions of parametric and statistical uncertainty. An approach to solving the problem based on NDS optimization using non-square cost functions is proposed. This approach, in addition to optimizing estimates, allows minimizing possible deviations of the NDS diagnostic parameters from tolerances. Adaptive procedures for protecting the EKF from divergence in the formation of estimates based on limiting the ratio of quadratic forms of output to input residuals of observations are also considered. Such procedures are based on the H-infinity theory. The two-point boundary value problems formulated for optimization and adaptation in the Euler-Lagrange form are solved by the invariant embedding method. The possibilities of the presented algorithms are discussed.

Keywords: nonlinear dynamical system, parameter observers, controlled error compensation, inverse problem of dynamics, non-quadratic optimization, adaptive estimation, H-infinity problem

1. Introduction

A dynamical system (DS) is one whose parameters change in space and time. The evolution of the state of such a system in time t can be described by a nonlinear differential equation of the form

$$dY(t)/dt = \dot{Y}(t) = F[Y(t)]; \tag{1}$$

where $Y(t)$ is a vector of state parameters of an unperturbed DS; $F(\ldots)$ is a matrix of variable coefficients characterizing the dynamics of changes in DS parameters.

In a real environment, the DS is affected by disturbances and noise, taking into account which the model of changes in the DS parameters will have the form

$$\dot{Y}_a(t) = F[Y_a(t)] + G(t)\xi(t), \tag{2}$$

where $Y_a(t)$ is a vector of actual state parameters; $\xi(t)$ is a disturbances vector; $G(t)$ is a matrix of coefficients characterizing the intensities of disturbances.

In the observable DS, the problem of restoring the state parameters of such a DS can be solved using external reference measurements,

$$Z^*(t) = h[Y(t)] + \vartheta(t), \tag{3}$$

where $\vartheta(t)$ is the l-dimensional vector of measurement errors.

However, in practical applications, it is more convenient to restore not the parameters of a nonlinear DC, but to determine their deviations from the reference phase trajectory. Such deviations correspond to errors in stabilization of the DS relative to the reference trajectory. The differential equation for the DS errors is obtained from the difference between relations (1) and (2) and has the form

$$dx(t)/dt = \dot{x}(t) = A(t)x(t) + G(t)\xi(t), \tag{4}$$

where $x(t) = \Delta Y(t) = Y_a(t) - Y(t)$ is the n-dimensional vector of actual DS errors; $A(t) = \partial F[Y(t)]/\partial Y|_{Y(t)=Y_a(t)}$ is a matrix of variable coefficients characterizing the dynamics of changes in DS errors.

Eq. (4) can be put into correspondence with the signals of observations of DS errors

$$Z(t) = Z^*(t) - Z^{**}(t), \tag{5}$$

and their predictive model

$$Z(t) = H(t)x(t) + \vartheta(t), \tag{6}$$

where $H(t) = \partial h[Y(t)]/\partial Y|_{Y(t)=Y_a(t)}$ is the matrix of the relationship of observations with the error vector DS; $Z^{**}(t) = h[(Y_a]$ is a parameter equivalent to the measurement of $h(y)$ and formed based on information from the actual DS.

Taking into account Eqs. (5) and (6), the problem of obtaining optimal estimates of the DS error vector using the least squares method can be formulated [1].

$$\hat{x}(t) = \arg\min_x J(t), \tag{7}$$

where $J(t) = \nu^T(t)R^{-1}(t)\nu(t)$; $\nu(t) = Z(t) - \hat{Z}(t)$ is a residual; $\hat{Z}(t) = H(t)\hat{x}(t)$ is the predicted value of the observation; ^ is the symbol for estimate; $E\left[\vartheta(t)\vartheta^{\mathrm{T}}(t-\tau)\right] = R(t)\delta(t-\tau)$ is a covariance matrix that normalizes the residuals; $\delta(t-\tau)$ is the delta function; E [...] is the operator of mathematical expectation.

The solution to problem (7) has the form

$$\hat{x}(t) = \left[H^{\mathrm{T}}(t)R^{-1}H(t)\right]^{-1}H^{\mathrm{T}}R^{-1}Z(t). \tag{8}$$

It should be noted that for $l < n$, only an approximate estimate $\hat{x}(t)$ of the error vector can be obtained from Eq. (8). In addition, Eq. (8) does not explicitly take into account the dynamics of changes in DC parameters, errors, and disturbances. Therefore, there is a need to combine procedures (2)–(6) and (8) into a single structure based on the diagram presented in **Figure 1**, where EMS is an external measuring system; OPS is an observation processing system; K_x is a matrix of coefficients for controlled compensation of DS error estimates; LCE is the loop for compensating estimates of errors; and LUP is the loop for updating parameters.

The following equation of a traditional observer [2] can be taken as a basis for implementing the scheme in **Figure 1**

$$\dot{\hat{x}}(t) = A(t)\hat{x}(t) + K(t)\nu(t). \tag{9}$$

The modifications of the observers are related to the choice of the K gain factor. The gain factor, formed according to the criterion of minimum variance of errors in estimating the state vector of a nonlinear DS, corresponds to the extended Kalman filter (EKF) [3]. For such an estimation filter, the gain factor has the form

$$K(t) = P(t)H^T(t)R^{-1}(t), \tag{10}$$

where $P(t) = E\left[\delta(t)\delta^{\mathrm{T}}(t)\right]$ is the covariance matrix of estimation errors, which is determined from the solution of the Riccati equation

$$\dot{P}(t) = A(t)P(t) + P(t)A^T(t) - K(t)R(t)K^T(t); \tag{11}$$

$\delta(t) = x(t) - \hat{x}(t)$ is the vector of estimation errors.

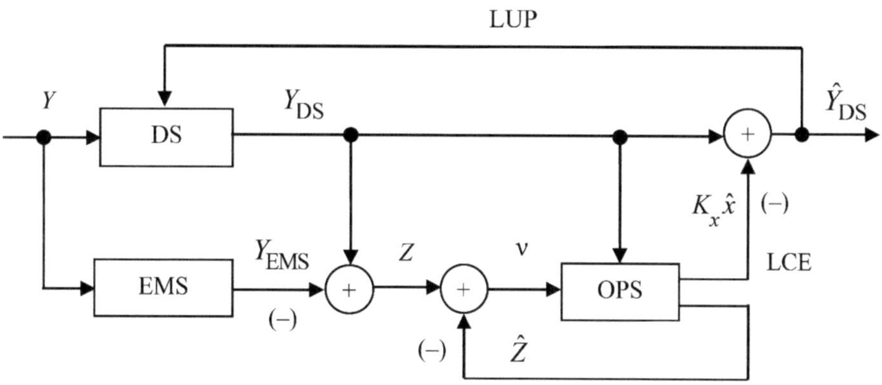

Figure 1.
Diagram of the observable dynamical system with estimation and error compensation loops.

It can be seen that in the traditional estimation algorithm, the gain factor does not depend on the observations. With parametric and statistical uncertainty, this leads to divergence of error estimates relative to their actual values. The correctness of the gain factor formation can be verified by mathematical modeling when the actual phase trajectory is known. If the DS model is adequate for the real process, then the actual estimation errors of the vector x should not exceed their root-mean-square values [4]. For example, for the jth element of the state vector x this means

$$\delta_j(t) < < \sigma_j = \sqrt{P_{jj}}. \tag{12}$$

It is believed that in stochastic systems, the non-divergent estimation process corresponds to the 3σ rule [5], that is,

$$\delta_j(t) < 3\sigma_j. \tag{13}$$

The implementation of the scheme in **Figure 1** assumes the use of error estimates in the DS. The following solutions to this problem are possible [3]:

- a scheme with a contour for compensation of estimates (see **Figure 1**, without a dashed line):

$$\dot{Y}(t) = F[Y(t)]; \tag{14}$$

$$\dot{\hat{x}}(t) = A(t)\hat{x}(t) + K(t)\nu(t); \tag{15}$$

$$\hat{Y}(t) = Y(t) - \hat{x}(t); \tag{16}$$

- a scheme with a contour for updating parameters (see **Figure 1**, with a dashed line):

$$\dot{\hat{Y}}(t) = F\left[\hat{Y}(t)\right] - \hat{x}(t) \tag{17}$$

$$\dot{\hat{x}}(t) = K(t)\nu(t). \tag{18}$$

In accordance with procedures (14)–(16), error estimates are written off from the parameters at the DS output without correcting Eq. (14). When implementing such a mode, the necessary conditions for the absence of divergence of the EKF (13), associated with the linearity of the change in the residual errors of the DS between observations, may not be fulfilled. Therefore, it seems appropriate to perform controlled compensation of DS error estimates between observation sessions. When implementing procedures (17) and (18), the parameters in DS (17) are corrected using error estimates with their subsequent zeroing. However, in this case, the error prediction module (4) is excluded from the DS structure. This module supports the smoothing properties of the EKF relative to anomalous observation noises. Therefore, it seems appropriate to perform controlled compensation of DS error estimates between observation sessions. Such control can be implemented using the K_x coefficients for damping the estimates (see **Figure 1**). The above coefficients should be determined in such a way as to minimize the errors of the estimates formed by the EKF relative to the reference observed parameters. It can be noted that the integration cycles of the DS equations, their errors, and the formation of observations may differ from each other. Next, approaches to solving the formulated problems will be considered.

2. Continuous estimation and controlled error compensation in nonlinear dynamical system

The mathematical description of the DS operation is based on the direct and inverse problem of dynamics. Stated classically, such problems can be given the following interpretation [6]:

- the essence of the direct problem: Based on the given initial conditions, input actions, and a function describing the dynamics of parameter changes, by solving Eqs. (2) and (4), the path of the DS in phase space is determined;
- the essence of the inverse problem: The control actions are determined that ensure the motion of the DS along a given path in phase space. For example, the solution to Eq. (4) is a matrix exponential. In this case, the inverse problem is reduced to identifying the coefficients $A(t)$ that ensure motion along such an exponential. However, the matrix $A(t)$ corresponds to the linearized Eq. (4). Therefore, the prediction errors of nonlinear DS are inaccurate. From here, the inverse problem is reduced to determining additional coefficients K_x that minimize the divergence between the reckoned $\hat{Y}(t)$ and reference $Y(t)$ paths formed from observations (3). The following modification of Eq. (17) corresponds to such a problem, that is,

$$\dot{Y}(t) = F\left[\hat{Y}(t)\right] - A(t)\hat{x}(t) + u(t), \tag{19}$$

where $u(t) = -K_x(t)\hat{x}(t)$ is a vector of controls that stabilize the DS on the reference path.

The matrix of coefficients $K_x(t)$ included in Eq. (19) corresponds to the exponential damping of the system errors between observations. It is assumed that $K_x(t) = \Lambda(t)$, where $\Lambda(t) = diag[\alpha_1(t)\cdots\alpha_n(t)]$ is the diagonal matrix of error damping coefficients; $\alpha_j(t) > 0$.

In the realization of Eq. (19), the problem of determining the estimates $\hat{\overline{\Lambda}}(t) = [\hat{\alpha}_1(t), \cdots\hat{\alpha}_n(t)]^{\mathrm{T}}$ of the vector of coefficients, which are optimal ones according to the appropriate criterion, still remains topical.

A similar problem in discrete time was considered in the work of Chernodarov [6]. This section presents procedures for identifying damping coefficients for DS error estimates in continuous and discrete-continuous time.

The problem of optimizing the procedures for damping DS errors can be solved on the basis of a united cost function, which combines the following parameters into a single structure:

a. residuals for DS error estimates;

b. residuals for estimates of error damping coefficients.

The next modification of observation (5) and its residual correspond to the scheme with the correction of the DS parameters based on the error estimates

$$Z(t) = h[Y(t), t]|_{Y(t)=Y_a(t)} - h[Y(t), t]. \tag{20}$$

Taking into account (17), we have

$$\nu(t) = Z(t). \tag{21}$$

Optimization problem (a) with respect to the estimates of the DS state parameters can be solved on the basis of the quadratic cost function

$$J(t) = 0.5 \int_{t_0}^{t_f} \left\{ \|\nu(t)\|^2_{R^{-1}(t)} + \|\xi(t)\|^2_{Q^{-1}(t)} \right\} d\tau \rightarrow \min_{Y(t)} \tag{22}$$

taking into account the constraint on the dynamics of change in the system parameters

$$\dot{Y}(t) = F(Y, t) + G(T)\xi(t), \tag{23}$$

where $T = [t_0, t_f]$ is the optimization interval; $E[\xi(t)\xi^T(t - \tau)] = Q(t)\delta(t - \tau)$ is a covariance matrix for the disturbance vector; $\|\xi(t)\|^2_{Q^{-1}} = \xi^{\mathrm{T}}(t)Q^{-1}\xi(t)$.

The solution to problem (22) under constraint (23) can be obtained using R. Sridhar's technique [7, 8], which is based on the Pontryagin minimum principle [9] and on the method of invariant imbedding [10]. This technique allows us to move from conditional (22), (23) to unconditional optimization using the Hamilton function

$$\mathrm{H} = 0.5\left\{ \|\nu(t)\|^2_{R^{-1}(t)} + \|\xi\|^2_{Q^{-1}(t)} \right\} + \lambda_Y^T(t)[F(Y, t) + G(t)\xi(t)], \tag{24}$$

and on the solution of the canonical equations

$$\dot{\hat{Y}}(t) = \frac{\partial \mathrm{H}}{\partial \lambda_Y^T(t)}; \dot{\hat{\lambda}}_Y^T(t) = -\frac{\partial \mathrm{H}}{\partial Y(t)}; \frac{\partial \mathrm{H}}{\partial \xi(t)} = 0 \tag{25}$$

with the boundary conditions

$$\hat{\lambda}_Y(t_0) = \mathrm{var}; \hat{\lambda}_Y(t_f) = 0; \hat{Y}(t_0) = 0; \hat{Y}(t_f) = \mathrm{var}, \tag{26}$$

where $\lambda_Y(t)$ is the vector of Lagrange multipliers. The introduction of the Lagrange multiplier vector [11] into Eq. (24) allows us to move from the problem of minimizing the functional (22) with constraints (23) to the problem of an unconditional minimum.

By calculating partial derivatives and performing algebraic transformations, we obtain an expanded notation of the canonical equations

$$\dot{Y}(t) = F(Y, t) - G(t)Q(t)G(t)\lambda_Y(t) = \gamma[Y(t), \lambda_Y(t), t]; \tag{27}$$

$$\dot{\lambda}_Y(t) = -H^{\mathrm{T}}(t)R^{-1}(t)\nu(t) - A^{\mathrm{T}}(t)\lambda_Y(t) = \mu[Y(t), \lambda_Y(t), t]; \tag{28}$$

$$\xi(t) = -Q(t)G^{\mathrm{T}}(t)\lambda_Y(t). \tag{29}$$

To find the $\dot{\hat{Y}}(t)$ estimate of the DS parameter vector from the observations of $Z(t)$, it is necessary to solve the two-point boundary value problem (TPBP), which is determined by Eqs. (27)–(29) and boundary conditions (26). The solution of the above TPBP can be found by the method of invariant imbedding [10]. According to this method, TPBP is included in a more general problem with the following boundary conditions

$$\hat{\lambda}_Y(t_f) = c; \hat{Y}(t_f) = r(c, t_f), \tag{30}$$

It can be shown that the dynamics of change in the $r(c, t_f)$ parameter can be described by the following Eq. (10)

$$\frac{\partial r(c, t_f)}{\partial t_f} + \frac{\partial r(c, t_f)}{\partial c} \mu[r(c, t_f), c, t_f] = \gamma[r(c, t_f), t_f]. \tag{31}$$

The solution of Eq. (31) can be approximated by a linear function

$$r(c, t_f) = \hat{Y}(t_f) - P(t_f)c, \tag{32}$$

where $P(t_f)$ and c are, respectively, the matrix of coefficients and the vector of uncertain multipliers that determine the deviations of estimates from the true values of the parameters. Here, the subtraction sign corresponds to the write-off of error estimates from the parameters.

If we substitute (32) into (31) and consider the time t_f as the current t, we obtain

$$\dot{\hat{Y}}(t) - \dot{P}(t)c - P(t)\,\mu[\hat{Y}(t) - P(t)c, c, t] = \gamma[\hat{Y}(t) - P(t)c, c, t]. \tag{33}$$

The solution to Eq. (33) can be obtained by expanding the parameters γ and μ in a Taylor series with respect to Y and restricting ourselves to linear terms

$$\begin{aligned} &\dot{\hat{Y}}(t) - \dot{P}(t)c - P(t)\,\mu\,[\hat{Y}(t), c, t] + P(t)\frac{\partial \mu[\hat{Y}(t), c, t]}{\partial Y}P(t)c = \\ &= \gamma\,[\hat{Y}(t), c, t] - \frac{\partial \gamma(\hat{Y}, c, t)}{\partial Y}P(t)c. \end{aligned} \tag{34}$$

Substituting into (34) the parameters γ and μ, as well as the partial derivatives with respect to Y from Eqs. (27) and (28), we obtain

$$\begin{aligned} &F(Y,t) - G(t)Q(t)G^{\mathrm{T}}(t)c - A(t)\,P(t)c = P(t)H^{\mathrm{T}}(t)R^{-1}(t)\nu(t) + \\ &+ P(t)A^{\mathrm{T}}(t)c - P(t)H^{\mathrm{T}}(t)R^{-1}H(t)P(t)c + \dot{Y}(t) - \dot{P}(t)c, \end{aligned} \tag{35}$$

where matrices $G(t), A(t), H(t), R(t), P(t)$, and $Q(t)$ have expressions presented in Eqs. (2), (4), (6), (8), (10), and (23), respectively.

By equating in (35) the terms that include and do not include the parameter c, we obtain differential Eqs. (17), (18) for the estimates and (11) for the covariance matrix $P(t)$.

The technique for obtaining expressions (11), (17), (18) taking into account (32)–(35) can be presented in a more compact form

$$\dot{\hat{Y}}(t) = \gamma[\hat{Y}(t), c, t]\big|_{c=0} + P(t)\mu[\hat{Y}(t), c, t]\big|_{c=0}; \tag{36}$$

$$\dot{P}(t) = -\left.\frac{\partial\gamma\left[\hat{Y}(t),c,t\right]}{\partial c}\right|_{c=0} - P(t)\left.\frac{\partial\mu\left[\hat{Y}(t),c,t\right]}{\partial c}\right|_{c=0}. \tag{37}$$

Using the technique (24)–(37), it is possible to solve the problem (19) or (b) of determining the damping coefficients of the DS error estimates between the moments of observations. For this purpose, by analogy with (24), the Hamilton function of the following form can be used

$$\begin{gathered}\mathrm{H} = 0.5\left\{\|\nu(t)\|^2_{R^{-1}(t)} + \|\xi\|^2_{Q^{-1}(t)} + \|\eta(t)\|^2_{P_Y^{-1}(t)}\right\} + \\ +\lambda_Y^T(t)[F(Y,t) - \Lambda(t)Y(t) + G(t)\xi(t)] + \lambda_\Lambda^T(t)B(t)\overline{\Lambda}(t)\end{gathered}, \tag{38}$$

where

$$\eta(t) = \Lambda(t)Y(t) - K(t)\nu(t) = \tilde{Y}(t)\overline{\Lambda}(t) - K(t)\nu(t) \tag{39}$$

is the residuals between the predicted and observed corrections for the estimates of the vector of parameters of the DS; $\lambda_\Lambda(t)$ is the vector of Lagrange multipliers; $B(t)$ is a matrix of gain factors characterizing the dynamics of changes in the damping coefficients of estimates between observation sessions;
$\tilde{Y}(t) = diag[Y_1(t)\,...\,Y_n(t)]; \hat{\overline{\Lambda}}(t) = [\hat{\alpha}_1(t),\cdots\hat{\alpha}_n(t)]^{\mathrm{T}}$; $\Lambda(t) = diag[(\alpha_1(t)\cdots\alpha_n(t)]$.

By analogy with (25) and (26), canonical equations can be formed

$$\dot{Y}(t) = \frac{\partial\mathrm{H}}{\partial\lambda_Y^T(t)}; \dot{\lambda}_Y^T(t) = -\frac{\partial\mathrm{H}}{\partial Y(t)}; \frac{\partial\mathrm{H}}{\partial\xi(t)} = 0; \tag{40}$$

$$\dot{\overline{\Lambda}}(t) = \frac{\partial\mathrm{H}}{\partial\lambda_\Lambda^T(t)}; \dot{\lambda}_\Lambda^T(t) = -\frac{\partial\mathrm{H}}{\partial\overline{\Lambda}(t)}; \tag{41}$$

$$\lambda_Y(t_0) = \mathrm{var}; \lambda_Y(t_f) = 0; \hat{Y}(t_0) = 0; \hat{Y}(t_f) = \mathrm{var}; \tag{42}$$

$$\lambda_\Lambda(t_0) = \mathrm{var}; \lambda_\Lambda(t_f) = 0; \hat{\Lambda}(t_0) = 0; \hat{\Lambda}(t_f) = \mathrm{var}. \tag{43}$$

By calculating partial derivatives and performing algebraic transformations, we obtain an expanded notation of the canonical equations

$$\dot{Y}(t) = F(Y,t) - \Lambda(t)Y(t) - G(t)Q(t)G(t)\lambda_Y(t) = \gamma_Y[Y(t),\Lambda(t),\lambda_Y(t)] \tag{44}$$

$$\dot{\overline{\Lambda}}(t) = B(t)\overline{\Lambda}(t) = \gamma_\Lambda\left[\overline{\Lambda}(t),t\right]; \tag{45}$$

$$\begin{gathered}\dot{\lambda}_Y(t) = -H^{\mathrm{T}}(t)R^{-1}(t)\nu(t) + H^{\mathrm{T}}(t)K^{\mathrm{T}}(t)P_Y^{-1}(t)\eta(t) - \\ -\left[A^{\mathrm{T}}(t) - \Lambda^{\mathrm{T}}(t)\right]\lambda_Y(t) = \mu_Y\left[Y(t),\Lambda(t),\lambda_y(t),t\right];\end{gathered} \tag{46}$$

$$\begin{gathered}\dot{\lambda}_\Lambda(t) = \tilde{Y}^{\mathrm{T}}(t)\lambda_Y(t) - \tilde{Y}^{\mathrm{T}}(t)P_Y^{-1}(t)\eta(t) - B^{\mathrm{T}}(t)\lambda_\Lambda(t) = \\ = \mu_\Lambda[Y(t),\Lambda(t),\lambda_Y,\lambda_\Lambda,t];\end{gathered} \tag{47}$$

$$\xi(t) = -Q(t)G^{\mathrm{T}}(t)\lambda_Y(t). \tag{48}$$

Let us introduce notation for parameters

$$\lambda(t) = \begin{bmatrix}\lambda_Y(t)\\ \lambda_\Lambda(t)\end{bmatrix}; y(t) = \begin{bmatrix}Y(t)\\ \overline{\Lambda}(t)\end{bmatrix} \tag{49}$$

and solutions

$$\gamma(Y,\Lambda,C,t)=\begin{bmatrix}\gamma_Y(t)\\ \gamma_\Lambda(t)\end{bmatrix};\mu(Y,\Lambda,C,t)=\begin{bmatrix}\mu_Y(t)\\ \mu_\Lambda(t)\end{bmatrix}. \tag{50}$$

The TPBP (44)–(48) can be solved by taking into account the expansion of the boundary conditions (42) and (43) by analogy with (30) and (32)

$$y(t_f)=r(C,t_f)=\begin{bmatrix}r_Y(C,t_f)\\ r_\Lambda(C,t_f)\end{bmatrix}; \tag{51}$$

$$\lambda(t_f)=C=\begin{bmatrix}c_Y\\ c_\Lambda\end{bmatrix}. \tag{52}$$

By analogy with (31) for conditions (51) and (52), the basic equation of invariant imbedding is valid

$$\frac{\partial r}{\partial t_f}+\frac{\partial r}{\partial C}\mu(r,C,t_f)=\gamma(r,C,t_f). \tag{53}$$

The solution of Eq. (53), by analogy with (32), can be represented in general form by the following expression

$$r(C,t_f)=\hat{y}(t_f)-P(t_f)C, \tag{54}$$

where $P(t_f)=\begin{bmatrix}P_Y(t_f) & P_{Y\Lambda}(t_f)\\ P_{\Lambda Y}(t_f) & P_\Lambda(t_f)\end{bmatrix}$ is a matrix of coefficients that determine the spread of estimates relative to the true values of the parameters. This may be a covariance matrix of estimate errors.

Considering (44)–(54) and assuming the time t_f to be the current t, the TPBP solution according to the algorithm (36) and (37) will have the form:

$$\begin{bmatrix}\dot{Y}(t)\\ \dot{\Lambda}(t)\end{bmatrix}=\begin{bmatrix}\gamma_Y(t)\\ \gamma_\Lambda(t)\end{bmatrix}_{C=0}+\begin{bmatrix}P_Y(t) & P_{Y\Lambda}(t)\\ P_{\Lambda Y}(t) & P_\Lambda(t)\end{bmatrix}\begin{bmatrix}\mu_Y(t)\\ \mu_\Lambda(t)\end{bmatrix}_{C=0}; \tag{55}$$

$$\begin{bmatrix}\dot{P}_Y(t) & \dot{P}_{Y\Lambda}(t)\\ \dot{P}_{\Lambda Y}(t) & \dot{P}_\Lambda(t)\end{bmatrix}=-\begin{bmatrix}\frac{\partial\gamma_Y}{\partial c_Y} & \frac{\partial\gamma_Y}{\partial c_\Lambda}\\ \frac{\partial\gamma_\Lambda}{\partial c_Y} & \frac{\partial\gamma_\Lambda}{\partial c_\Lambda}\end{bmatrix}_{C=0}-\begin{bmatrix}P_Y(t) & P_{Y\Lambda}(t)\\ P_{\Lambda Y}(t) & P_\Lambda(t)\end{bmatrix}\begin{bmatrix}\frac{\partial\mu_Y}{\partial c_Y} & \frac{\partial\mu_Y}{\partial c_\Lambda}\\ \frac{\partial\mu_\Lambda}{\partial c_Y} & \frac{\partial\mu_\Lambda}{\partial c_\Lambda}\end{bmatrix}_{C=0}. \tag{56}$$

The solution of Eqs. (55) and (56) has the form:

$$\dot{\hat{Y}}(t)=F(\hat{Y},t)-\hat{\Lambda}(t)\hat{Y}(t)-K_Y(t)\nu(t)+K_{Y\eta}(t)\eta(t)-K_{Y\Lambda}(t)\eta(t); \tag{57}$$

$$\dot{\hat{\bar{\Lambda}}}(t)=B(t)\hat{\bar{\Lambda}}(t)-K_{\Lambda Y}(t)\nu(t)+K_{\Lambda\eta}(t)\eta(t)-K_\Lambda(t)\eta(t); \tag{58}$$

$$\begin{aligned}\dot{P}_Y(t)&=A(t)P_Y(t)+P_Y(t)A^T(t)+G(t)Q(t)G^T(t)-K_Y(t)H(t)P_Y(t)-\\&-K_{Y\eta}H(t)P_Y(t)+K_{Y\eta}(t)\hat{\Lambda}(t)P_Y-\hat{\Lambda}(t)P_Y(t)-P_Y(t)\hat{\Lambda}^T(t)-P_{Y\Lambda}(t)\dot{\hat{Y}}^T+\\&+P_{Y\Lambda}(t)B^T(t)+K_{Y\Lambda}(t)\Lambda(t)P_Y(t)-K_{Y\Lambda}(t)H(t)P_Y(t);\end{aligned} \tag{59}$$

$$\dot{P}_{\Lambda}(t) = B(t)P_{\Lambda}(t) + P_{\Lambda}(t)B^{\mathrm{T}}(t) + K_{\Lambda\eta}(t)\hat{\hat{Y}}^{\mathrm{T}}(t)P_Y(t) - K_{\Lambda}(t)\hat{\hat{Y}}(t)P_{\Lambda}(t); \qquad (60)$$

$$\dot{P}_{Y\Lambda}(t) = \hat{\hat{Y}}(t)P_{\Lambda} + P_{Y\Lambda}(t)B^{\mathrm{T}}(t) - K_{Y\Lambda}(t)\hat{\hat{Y}}(t)P_{\Lambda}(t) + K_{Y\eta}(t)\hat{\hat{Y}}(t)P_{\Lambda}(t); \qquad (61)$$

$$\begin{aligned}\dot{P}_{\Lambda Y}(t) &= P_{\Lambda Y}(t)A^{\mathrm{T}}(t) - P_{\Lambda Y}(t)\hat{\Lambda}^{\mathrm{T}}(t) - P_{\Lambda}(t)\hat{\hat{Y}}^{\mathrm{T}}(t) + \\ &+ K_{\Lambda\eta}(t)H(t)P_Y(t) - K_{\Lambda Y}(t)H(t)P_Y(t) - K_{\Lambda}(t)H(t)P_Y;\end{aligned} \qquad (62)$$

where $K_Y(t) = P_Y(t)H^{\mathrm{T}}(t)R^{-1}(t)$; $K_{Y\eta}(t) = P_Y(t)H^{\mathrm{T}}(t)K^{\mathrm{T}}(t)P_Y^{-1}(t)$;
$K_{Y\Lambda}(t) = P_{Y\Lambda}(t)\hat{\hat{Y}}^{\mathrm{T}}(t){P_Y}^{-1}(t)$; $K_{\Lambda\eta}(t) = P_{\Lambda Y}(t)H^{\mathrm{T}}(t)K^{\mathrm{T}}(t)P_Y^{-1}(t)$;
$K_{\Lambda Y}(t) = P_{\Lambda Y}(t)H^{\mathrm{T}}(t)R^{-1}(t)$; $K_{\Lambda}(t) = P_{\Lambda}(t)\hat{\hat{Y}}^{\mathrm{T}}(t){P_Y}^{-1}(t)$.

Assuming that the parameters $Y(t)$ and $\overline{\Lambda}(t)$ are independent, Eqs. (57)–(62) take the form

$$\dot{\hat{Y}}(t) = F(\hat{Y}, t) - \hat{\Lambda}(t)\hat{Y}(t) - K_Y(t)\nu(t) + K_{Y\eta}(t)\eta(t); \qquad (63)$$

$$\dot{\hat{\overline{\Lambda}}}(t) = B(t)\hat{\overline{\Lambda}}(t) - K_{\Lambda}(t)\eta(t); \qquad (64)$$

$$\begin{aligned}\dot{P}_Y(t) &= A(t)P_Y(t) + P_Y(t)A^{\mathrm{T}}(t) + G(t)Q(t)G^{\mathrm{T}}(t) - K_Y(t)H(t)P_Y(t) - \\ &- K_{Y\eta}H(t)P_Y(t) + K_{Y\eta}(t)\hat{\Lambda}(t)P_Y - \hat{\Lambda}(t)P_Y(t) - P_Y(t)\hat{\Lambda}^{\mathrm{T}}(t);\end{aligned} \qquad (65)$$

$$\dot{P}_{\Lambda}(t) = B(t)P_{\Lambda}(t) + P_{\Lambda}(t)B^{\mathrm{T}}(t) - K_{\Lambda}(t)\hat{\hat{Y}}(t)P_{\Lambda}(t). \qquad (66)$$

When implementing Eqs. (63)–(66), it is assumed that observations are formed continuously and synchronously with the integration step. However, observations may arrive at discrete points in time that do not coincide with the integration step. In this case, the error equations and the corresponding damping coefficients can be formed on the basis of a discrete modification of the EKF.

3. Discrete estimation and continuous error compensation in nonlinear dynamical system

The problem of formation and controlled compensation of estimates $\hat{x}_i = \hat{x}(t_i)$ of the error vector x_i of a nonlinear DS (2) based on observations Z_i obtained at discrete time points t_i is considered. A discrete analogue of the error Eq. (4) corresponds to such observations

$$x_i = \Phi_i x_{i-1} + \Gamma_i \xi_{i-1}, \qquad (67)$$

where Φ_i is the transition matrix for the vector of errors, and this matrix is determined from the solution of the differential equation

$$d\Phi(t, t_{i-1})/dt = \dot{\Phi}(t) = A(t)\Phi(t, t_{i-1}) \text{ for } \Phi(t_{i-1}, t_{i-1}) = I; \qquad (68)$$

I is an identity matrix of the appropriate dimension; Γ_i is the transition matrix for the disturbance vector ξ_i. The given matrix is the solution of the inhomogeneous part of Eq. (4)

$$d\Gamma(t, t_i)/dt = \dot{\Gamma}_i = A(t)\Gamma(t, t_{i-1}) + G(t); \Gamma(t_{i-1}, t_{i-1}) = 0. \qquad (69)$$

The following options are possible [3] for the relationship between the DS parameters and its error estimates in discrete time:

a. diagram with a loop for compensating for estimates (see **Figure 1**, without a dashed line):

$$\hat{x}_{i/i-1} = \Phi_i \hat{x}_{i-1/i-1}; \tag{70}$$

$$\hat{x}_{i/i} = \hat{x}_{i/i-1} + K_i\left(Z_i - H_i\hat{x}_{i/i-1}\right); \hat{Y}_{i/i} = \hat{Y}_{i/i-1} - \hat{x}_{i/i}; \tag{71}$$

b. diagram with a loop for updating parameters (see **Figure 1**, with a dashed line):

$$\hat{x}_{i/i} = K_i Z_i; \hat{Y}_{i/i} = \hat{Y}_{i/i-1} - \hat{x}_{i/i}; \hat{x}_{i/i}{:=}0, \tag{72}$$

where $\hat{Y}_{i/i-1}$, $\hat{x}_{i/i-1}$ are the predicted values of estimates both of the vector of parameters and of the DS errors at the instant $t = t_i$ of time after Z_{i-1} observations have been processed; $\hat{Y}_{i/i}$, $\hat{x}_{i/i}$ are the updated values of estimates after Z_i observations have been processed; K_i is the EKF amplification factor.

For Gaussian perturbations, the implementation of scheme b) based on a sequential modification [12] of the discrete EKF has the following form

Prediction

$$m_0 = \hat{x}_{i/i-1} = \tilde{\Phi}_i \hat{x}_{i-1/i-1}; j = 1; \tag{73}$$

$$M_0 = P_{i/i-1} = \Phi_i P_{i-1/i-1} \Phi_i^T + \Gamma_i Q_{i-1} \Gamma_i \tag{74}$$

Updating:

$$\nu_j = z_j - H_j m_{j-1}; \tag{75}$$

$$\alpha_j^2 = H_j M_{j-1} H_j^{\mathrm{T}} + R_j; \tag{76}$$

$$K_j = M_{j-1} H_j^{\mathrm{T}} / \alpha_j^2; \tag{77}$$

$$m_j = m_{j-1} + K_j \nu_j; \tag{78}$$

$$M_j = \left(K_j H_j - I\right) M_{j-1} \left(K_j H_j - I\right)^{\mathrm{T}} - K_j R_j K_j^{\mathrm{T}}; \quad j = \overline{1,\ l}; \tag{79}$$

$$\hat{x}_{i/i} = m_l; P_{i/i} = M_l, \tag{80}$$

where $Z_i = Z(t_i) = \left[z_{1(i)} \ldots z_{j(i)} \ldots z_{l(i)}\right]^{\mathrm{T}}; H_i = \left[H_{1(i)}^{\mathrm{T}} \ldots H_{j(i)}^{\mathrm{T}} \ldots H_{l(i)}^{\mathrm{T}}\right]^{\mathrm{T}}$.

Traditionally [3], the processing of observations in the EKF ends with the procedure

$$\hat{Y}_{i/i} = \hat{Y}_{i/i-1} - \hat{x}_{i/i}, \text{ and further } \hat{x}_{i/i}{:=}0. \tag{81}$$

Following the approach considered in the paper, an incomplete compensation of DS error estimates after processing observations is proposed

$$\hat{Y}_{i/i} = \hat{Y}_{i/i-1} - \hat{\Lambda}_i \hat{x}_{i/i}, \tag{82}$$

where $\hat{\Lambda}_{ii}$ is a matrix for damping coefficients of estimates at the i-point in time.

For a continuous-time DS, the relationship of its parameters with error estimates generated from discrete observations has the form

$$\dot{\hat{Y}}(t) = F\left[\hat{Y}(t)\right] - \hat{\Lambda}(t, t_{i-1})\hat{x}(t, t_{i-1}). \tag{83}$$

Between observation sessions, Eq. (83) has the following numerical solution

$$\hat{Y}_i = F\left(\hat{Y}_{i-1}, \Delta t_i\right) - \tilde{\Phi}_i \hat{x}_{i-1}, \tag{84}$$

where

$$\tilde{\Phi}_i = \tilde{\Phi}(t, t_{i-1}) = e^{-\hat{\Lambda}_i \Delta t_i} \approx I - \hat{\Lambda}_{ii}\Delta t_i; \tag{85}$$

$\hat{\overline{\Phi}}_i = \left[\hat{\overline{\Phi}}_{11(i)}, \cdots \hat{\overline{\Phi}}_{nn(i)}\right]^{\mathrm{T}}$; $\tilde{\Phi}_i = diag\left[\tilde{\Phi}_{11(i)} \cdots \tilde{\Phi}_{nn(i)}\right]$; $\Delta t_i = t_i - t_{i-1}$ is the integration step.

Based on the algorithm (72)–(85), the problem of forming damping coefficients $\overline{\Lambda}_i$ for error estimates arises. We will assume that the parameters $\overline{\Lambda}_i$ and the DS errors x_i are independent.

The optimality criterion will be formed based on the residuals η_i between the predicted $\hat{x}_{i/i-1}$ and updated $\hat{x}_{i/i}$ estimates of the DS errors, that is,

$$\hat{\overline{\Lambda}}_i = \arg\min_{\overline{\Lambda}_i} 0.5 \sum_{i=i_0}^{i_f} \eta_i^T P_{i/i}^{-1} \eta_i, \tag{86}$$

provided that $\overline{\Lambda}_i = \overline{\Lambda}_{i-1}$, that is, $B = I$ in Eq. (45),where

$$\eta_i = \tilde{\Phi}_i \hat{x}_{i-1/i-1} - \hat{x}_{i/i} = \hat{x}_{i/i-1} - \hat{x}_{i/i}; \tag{87}$$

$P_{i/i} = P_{x(i/i)}$ is the covariance matrix of estimation errors, which is formed by the EKF at the ith instant of time from i observations; i_0, i_f are the subscripts for initial and final values, respectively, of an observation on the optimization interval $\left[i_0; i_f\right]$.

The solution of the problem (86) was obtained using Pontryagin's method [8, 10]. The application of this method is based on the formation of the Hamiltonian

$$\mathrm{H} = 0.5\left\{\|\eta_i\|^2_{P_{i/i}^{-1}} + \|\omega_i\|^2_{N_i^{-1}}\right\} + \mu_i^{\mathrm{T}}\left(\overline{\Lambda}_{i-1} + \omega_i\right) \tag{88}$$

and the solution of the canonical equations

$$\frac{\partial \mathrm{H}}{\partial \mu_i^{\mathrm{T}}} = \hat{\overline{\Lambda}}_i; \ \frac{\partial \mathrm{H}}{\partial \overline{\Lambda}_{i-1}} = \hat{\mu}_{i-1}^{\mathrm{T}}; \ \frac{\partial \mathrm{H}}{\partial \omega_i} = 0 \tag{89}$$

with the boundary conditions

$$\hat{\overline{\Lambda}}_{i_0} = 0; \hat{\overline{\Lambda}}_{i_f} = \mathrm{var}; \hat{\mu}_{i_f = 0}; \hat{\mu}_{i_0} = \mathrm{var}, \tag{90}$$

where μ_i is the vector of Lagrange multipliers; ω_i is the vector of disturbances that affect the compensation coefficients, which is characterized by the covariance matrix $E\left[\omega(t)\omega^{\mathrm{T}}(t-\tau)\right] = N(t)\delta(t-\tau)$.

By calculating partial derivatives and performing algebraic transformations, we obtain an expanded notation of the canonical equations

$$\hat{\Lambda}_{i/i-1} = a = \hat{\Lambda}_{i-1/i-1} - N_{i-1}\hat{\mu}_{i-1}; \tag{91}$$

$$\hat{\mu}_i = b = \hat{\mu}_{i-1} + \hat{\tilde{x}}_{i-1}^{\mathrm{T}} P_{i/i}^{-1}\eta_i; \tag{92}$$

$$\hat{\omega}_i = -N_{i-1}\hat{\mu}_{i-1}, \tag{93}$$

where $\hat{\tilde{x}}_{i-1} = diag\left\{\hat{x}_{1(i-1)}, \dots, \hat{x}_{n(i-1)}\right\}$.

The estimates of the $\hat{\overline{\Lambda}}_{i/i}$ damping coefficients updated by observations can be found by solving the two-point boundary value problem (91)–(93) taking into account the conditions (90). By analogy with the continuous case (27)–(29), the solution of TPBP can be found by the method of invariant imbedding [10]. Following this method, TPBP is included in a more general problem with non-zero boundary conditions

$$\hat{\mu}_{i_f} = c; \hat{x}_{i_f}\left[c, i_f\right] := x_{i_f} - S_{i_f}c, \tag{94}$$

where $\hat{x}_{i_f}$ is the solution of the TPBP for c = 0, S_{i_f} is a matrix of weighting coefficients; for example, it can be a covariance matrix.

Taking into account the conditions (94), a discrete variant of the general solution TPBP (36)–(37) will have the following form

$$\hat{\Lambda}_{i/i} = a\left(\hat{\Lambda}_{i/i-1} - S_{i/i-1}c; c; i-1\right)\Big|_{c=0} + S_{i/i}b\left(\hat{\Lambda}_{i/i-1} - S_{i/i-1c;c;i-1)}\right)\Big|_{c=0}; \tag{95}$$

$$S_{i/i} = -\left[\frac{\partial a\left(\hat{\Lambda}_{i/i-1} - S_{i/i-1}; c; i-1\right)}{\partial c}\right]\Bigg|_{c=0}\left[\frac{\partial b\left(\hat{\Lambda}_{i/i-1} - S_{i/i-1}c; \mathrm{c}; i-1\right.}{\partial c}\right]\Bigg|_{c=0}^{-1}. \tag{96}$$

The partial derivatives for parameters (91) and (92) in Eqs. (95) and (96) are determined taking into account relations (90) and (94) and have the form

$$\frac{\partial a}{\partial c} = -S_{i-1/i-1} - N_{i-1}; \tag{97}$$

$$\frac{\partial b}{\partial c} = I + \left(\tilde{x}_{i-1}^{T} P_{i/i}^{-1}\tilde{x}_{i-1}\right)\left(S_{i-i/i-1} + N_{i-1}\right), \tag{98}$$

where $S_{i-1/i-1}$ is a matrix of weight coefficients, which were obtained at the $(i\text{-}1)th$ instant of time, having regard to $i-1$ observations.

Upon substitution of relations (97) and (98) into Eqs. (95) and (96) and on performance of algebraic transformations with the use of a lemma on matrix inversion [13], we obtain the following algorithm for error estimation.

Prediction:

$$\hat{\overline{\Lambda}}_{i/i-1} = \hat{\overline{\Lambda}}_{i-1/i-1}; \tag{99}$$

$$S_{i/i-1} = S_{i-1/i-1} + N_i. \tag{100}$$

Updating:

$$\eta_i = \tilde{\Phi}_i \hat{x}_{i-1/i-1} - \hat{x}_{i/i} = \hat{x}_{i/i-1} - \hat{x}_{i/i}; \tag{101}$$

$$V_i = \hat{\hat{x}}_{i/i-1} S_{i/i-1} \hat{\hat{x}}^{\mathrm{T}}_{i/i-1} + P_{i/i}; \tag{102}$$

$$S_{i/i} = S_{i/i-1} - S_{i/i-1} \hat{\hat{x}}^{\mathrm{T}}_{i/i-1} V_i^{-1} \hat{\hat{x}}_{i/i-1} S_{i/i-1}; \tag{103}$$

$$\hat{\overline{\Lambda}}_{i/i} = \hat{\overline{\Lambda}}_{i/i-1} + S_{i/i} \hat{\hat{x}}^{T}_{i/i-1} P^{-1}_{i/i} \eta_i, \tag{104}$$

where $\hat{\hat{x}}_{i/i-1} = diag\{\hat{x}_{1(i/i-1)}, \ldots, \hat{x}_{n(i/i-1)}\}$; $\hat{\overline{\Lambda}}_{i/i-1}$; $\hat{\overline{\Lambda}}_{i-1/i-1}$ are, respectively, the predicted and updated (by the residual η_i) estimates of the vector of coefficients that characterize the values of compensated errors; $S_{i/i-1}$; $S_{i/i}$ are, respectively, the predicted and updated covariance matrices for the errors of forming the compensation coefficients, which are caused by dynamic noise environment.

In the implementation of the algorithm (99)–(104), the need arises for inversion of the matrices V_i and $P_{i/i}$ in Eqs. (103) and (104). These problems can be solved by assuming that the elements of vector $\overline{\Lambda}_i$ are mutually independent, and also considering only the diagonal elements of the matrix $P_{i/i}$. Such elements of the $P_{i/i}$ matrix are variances of errors in estimating the parameters of vector x_i. We will also assume that the dimensions of the vectors $\overline{\Lambda}_i$ and x_i coincide and are equal to $n \times 1$. In this case, by analogy with algorithm (73)–(80), a sequential implementation of the modules (99)–(104) is possible, namely:

Prediction:

$$\hat{\overline{\Lambda}}_0 = \hat{\overline{\Lambda}}_{i/i-1} = \hat{\overline{\Lambda}}_{i-1/i-1}; \tag{105}$$

$$S_0 = S_{i/i-1} = S_{i-1/i-1} + N_i. \tag{106}$$

Updating:

$$\eta_j = \hat{x}_{j(i/i-1)} - \hat{x}_{j(i/i)}; \tag{107}$$

$$V_j = \hat{\hat{x}}_{j(i)} S_{j-1} \hat{\hat{x}}^{\mathrm{T}}_{j(i)} + P_{jj(i/i)}; \tag{108}$$

$$\tilde{K}_j = S_{j-1} \hat{\hat{x}}^{\mathrm{T}}_{j(i)} / V_j; \tag{109}$$

$$S_j = \left(\tilde{K}_j \hat{\hat{x}}_{j(i)} - I\right) S_{j-1} \left(\tilde{K}_j \hat{\hat{x}}_{j(i)} - I\right)^{\mathrm{T}} - \tilde{K}_j P_{jj(i)} \tilde{K}_j^{\mathrm{T}}; \tag{110}$$

$$\hat{\overline{\Lambda}}_j = \hat{\overline{\Lambda}}_{j-1} + K_j \eta_j; j = \overline{1, n}, \tag{111}$$

where $\hat{\hat{x}}_i = \left[\hat{\hat{x}}^{\mathrm{T}}_{1(i)} \cdots \hat{\hat{x}}^{\mathrm{T}}_{j(i)} \cdots \hat{\hat{x}}^{\mathrm{T}}_{n(i)}\right]^{\mathrm{T}}$; $P_{jj(i/i)}$ is the jth diagonal element of the $P_{i/i}$ matrix; $\hat{x}_{j(i/i-1)}$ and $\hat{x}_{j(i/i)}$ are the jth elements of vectors $\hat{x}_{i/i-1}$ and $\hat{x}_{i/i}$.

The initial values of the elements of the covariance matrix S_0 and the vector of coefficients $\hat{\overline{\Lambda}}_0$ are formed taking into account the dynamic properties of the DS, as well as the intervals and noise of observations.

4. Optimization of observations based on non-quadratic criteria

The practical application of EKF (73)–(81) is associated with the problem of ensuring the reliability of the obtained estimates, that is, with the fulfillment of conditions (12), (13). If the errors $\delta(t) = x(t) - \hat{x}(t)$ for the estimates $\hat{x}(t)$ were known, then such a problem would be solved trivially. However, the actual errors of the estimates in the ratios (12), (13) are known only in mathematical modeling, when the true values of the DS parameters are known. Therefore, the problem is solved through parameters and procedures indirectly related to estimation errors. Traditional procedures for adapting [14] EKF to real operating conditions are usually based on matching the predicted variances (76) for the residuals (75) with their values calculated from observations. It should be noted that the application of this approach involves the formation of decisive rules that determine the need to coordinate the specified parameters. This section discusses approaches that combine in a single structure the decisive rules and procedures for updating the covariance matrix of estimation errors based on observation results. For this purpose, it is proposed to perform multi-level optimization of estimates using non-quadratic criteria, including observations and diagnostic parameters for them. At the first level, estimates can be optimized for accuracy, and at the second level, for their reliability. The problem is to choose a generalized DS state parameter that satisfies the specified capabilities. Here, the normalized residual $\beta_j = \nu_j/\alpha_j$ is considered as such a parameter, where ν_j is the residual (75); α_j is the scale parameter. The scale parameter can be the mean square deviation $\sqrt{R_j}$ for the observation error. A similar approach to the optimization of estimates is formulated in [15, 16]. This section presents a solution to such a problem in an expanded form.

To resolve the EKF divergence problem, we will take the algorithm (73)–(81) as a basis and consider the multi-level optimization criterion, which is as follows:

$$\hat{x}_i = \arg\min_{x_i} \frac{1}{k}\sum_{i=i_0}^{i_f} d_i^k, \tag{112}$$

in view of the constraint equation,

$$\Phi_i x_{i-1} + \Gamma_i \xi_{i-1} - x_i = 0, \tag{113}$$

where $d_i = 0.5\left(\beta_i^2 - \gamma_1^2\right)$ is the residual between the generalized state parameter and its tolerance γ_1^2; $k > 1$.

It is assumed that functional (112) is used to form estimates when the generalized parameter exceeds the tolerance, that is, when $d_i > 0$ and $k > 0$. Under nominal conditions, when $d_i \leq 0$, the traditional EKF (73)–(81) should function.

The solution to the problem (112) is based on the statistical properties of the correct residuals ν_j and β_j. The relationship between such residuals and the absence of divergence of the EKF is determined by the following conditions [17, 18]:

$$\nu_j \in N\left(0, \alpha_j^2\right) \tag{114}$$

and

$$\beta_j^2 \in \chi^2(1, 2). \tag{115}$$

The rule of "three sigma" (3σ) [5] and the quantile *a* (1) = 0.02 [19] enables us to interpret condition (13) as the confidence interval

$$\beta_j^2 \leq \gamma_1^2 = E\left[\beta_j^2\right] + 3\sqrt{V\left[\beta_j^2\right]} = 1 + 3\sqrt{2} \approx 5.2 \tag{116}$$

and to implement a guaranteed-protection loop intended to prevent the EKF from being diverged. Here, V is the variance operator. The above loop is to be included as a component of the EKF structure in circumstances where the generalized parameter β_j^2 falls outside the tolerance limits, namely, if

$$d_j > = 0. \tag{117}$$

The solution to the problem (112), taking into account constraint (113), can be obtained using Pontryagin's method [8, 10] by forming the Hamiltonian

$$\mathrm{H} = \frac{1}{k} d_i^k + 0.5\|\xi_{i-1}\|^2_{Q_i^{-1}} + \lambda_i^{\mathrm{T}}(\Phi_i x_{i-1} + \Gamma_i \xi_{i-1}) \tag{118}$$

and on the solution of the canonical equations

$$\frac{\partial \mathrm{H}}{\partial \lambda_i^{\mathrm{T}}} = \hat{x}_i; \frac{\partial \mathrm{H}}{\partial x_{i-1}} = \hat{\lambda}_{i-1}^{\mathrm{T}}; \frac{\partial \mathrm{H}}{\partial \xi_i} = 0 \tag{119}$$

with the boundary conditions

$$\hat{x}_{i_0} = 0; \hat{x}_{i_f} = \mathrm{var}; \hat{\lambda}_{i_0} = \mathrm{var}\ \hat{\lambda}_{i_f} = 0, \tag{120}$$

where λ_i is the vector of Lagrange multipliers.

By calculating partial derivatives and performing algebraic transformations, we obtain an expanded notation of the canonical equations

$$\hat{x}_{i/i-1} = a = \Phi_i \hat{x}_{i-1/i-1} - \Gamma_i Q_{i-1} \Gamma_i^{\mathrm{T}} \Phi_i^{-\mathrm{T}} \lambda_{i-1}; \tag{121}$$

$$\hat{\lambda}_i = b = \Phi_i^{-\mathrm{T}} \hat{\lambda}_{i-1} + H_i^{\mathrm{T}} \alpha_i^{-1} \beta_i d_i^{k-1}; \tag{122}$$

$$\hat{\xi}_i = -Q_{i-1} \Gamma_i^{\mathrm{T}} \Phi_i^{-\mathrm{T}} \hat{\lambda}_{i-1}, \tag{123}$$

where $\Phi_i^{-\mathrm{T}} = \left(\Phi_i^{-1}\right)^{\mathrm{T}}$.

For finding the estimate $\hat{x}_{i/i}$ of the vector of errors at the ith instant of time from i observations, we need to solve only a two-point boundary problem, which is determined by Eqs. (121)–(123) and boundary conditions (120). The solution of the above TPBP can be found by the invariant embedding method represented by Eqs. (94)–(96), where the parameters $\hat{\Lambda}_{i/i}$ and $S_{i/i}$ are replaced by the error vector $\hat{x}_{i/i}$ and its covariance matrix $P_{i/i}$. In problem (112), the partial derivatives that enter into Eqs. (95) and (96) are determined from relations (121) and (122), taking into account expression (120) and assumptions (94), that is,

$$\frac{\partial a}{\partial c} = -\Phi_i P_{i-1/i-1} - \Gamma_i Q_{i-1} \Gamma_i^{\mathrm{T}} \Phi_i^{-\mathrm{T}}; \tag{124}$$

$$\frac{\partial b}{\partial c} = \Phi_i^{-\mathrm{T}} + H_i^{\mathrm{T}}\alpha_i^{-2}\mu_i H_i\left(\Phi_i P_{i-1/i-1} + \Gamma_i Q_{i-1}\Gamma_i^{\mathrm{T}}\Phi_i^{-\mathrm{T}}\right), \tag{125}$$

where $\mu_i = (k-1)d_i^{k-2}\beta_i^2 + d_i^{k-1}$; $P_{i-1/i-1}$ is the covariance matrix, which was obtained at the (i-1)th instant of time, having regard to $i-1$ observations.

Upon substitution of relations (124) and (125) into Eqs. (95) and (96) and performing algebraic transformations with the use of a lemma on matrix inversion [13], we obtain the adaptive algorithm for error estimation.

Prediction: (73)–(74).

Tuning:

$$\nu_j = z_j - H_j m_{j-1}; \tag{126}$$

$$\beta_j^2 = \nu_j^2/\alpha_j^2; \tag{127}$$

$$d_j = 0.5\left(\beta_j^2 - \gamma^2\right) \tag{128}$$

Integrity monitoring:

$$\text{if } d_j \le 0 \text{ then } \mu_j = 1 \text{ and k} = 1; \tag{129}$$

$$\text{if } d_j > 0 \text{ then } \mu_j = (k-1)d_j^{k-2}\beta_j^2 + d_j^{k-1}. \tag{130}$$

Updating:

$$K_j = M_{j-1}H_j^{\mathrm{T}}/\left(H_j M_{j-1}H_j^{\mathrm{T}}\mu_j + \alpha_j^2\right); \tag{131}$$

$$m_j = m_{j-1} + K_j d_j^{k-1}\nu_j; \tag{132}$$

$$M_j = \left(K_j H_j\mu_j - I\right)M_{j-1}\left(K_j H_j\mu_j - I\right)^T + K_j\alpha_j^2\mu_j K_j^T; \tag{133}$$

$$\hat{x}_{i/i} = m_l; P_{i/i} = M_l; j = \overline{1, l}, \tag{134}$$

where l is the dimension of the observation vector Z.

The value of the parameter k in Eqs. (130)–(133) determines the rate of convergence of the residuals to the tolerance γ_1^2. The parameter d_j ensures that the DS is tuned to the real observation process, taking into account the parrying of parametric uncertainty, as well as possible deviations of disturbances from the Gaussian distribution. This possibility is related to the fact that, unlike the traditional EKF, in the algorithm (126)–(134), the gain factor K_j and the covariance matrix M_j depend on the residuals ν_j for the observations.

Non-Quadratic Optimization of Observations Based on An H_∞ Criteria.

The reliability of the DS error estimates can be improved by performing the EKF output checking. Such a problem can be solved by checking the following norm of the transfer function $H(p)$ from perturbations of the DS ξ_i and observations ϑ_i to estimation errors δ_i

$$\|H(p)\|_\infty = \sup_{\xi,\,\vartheta\neq 0}\frac{\|\delta\|_2^2}{\|\xi\|_2^2 + \|\vartheta\|_2^2}, \tag{135}$$

where $\|\xi\|_2 = \left(\sum_{i=i_0}^{\infty} \xi_i^{\mathrm{T}} \xi_i \right)^{1/2}$ is the Euclidian norm for the vector ξ_i. The Euclidian norms for the vectors δ_i and ϑ_i are determined in a similar way. The specified disturbances and errors are presented in (3), (12), (22); sup denotes the upper bound.

For stable estimation, the following test condition can be formed:

$$\|H(p)\|_\infty \leq \gamma^2, \tag{136}$$

where $\gamma^2 \geq 0$is the parameter that can be considered as a tolerance.

Traditionally, it is believed [20] that in order to implement conditions (135) and (136), the exact observations associated with the errors must be known. Such observations are generally unknown. In this regard, it is possible to form an analogue of expression (135), in which the errors under consideration are related to observations. Such an analogue can be formed on the basis of the ratio of a priori and a posteriori residuals of observations. Then, criterion (135) can be associated with its following modification

$$\frac{\gamma^{-2} \sum_{i=i_0+1}^{i_f} \left\| \eta_{i/i} \right\|_2^2}{\sum_{i=i_0+1}^{i_f} \left(\|\xi_i\|_2^2 + \left\| \nu_{i/i-1} \right\|_2^2 \right)} \leq 1, \tag{137}$$

where

$$\nu_{i/i-1} = z_i - H_i \hat{x}_{i/i-1}; \tag{138}$$

$$\eta_{i/i} = z_i - H_i \hat{x}_{i/i}. \tag{139}$$

The mutual correspondence between criteria (135) and (137) can be seen if we write down the models for observations (138) and (139), namely:

$$\nu_{i/i-1} = H_i x_i + \vartheta_i - H_i \hat{x}_{i/i-1} = H_i \delta_{i/i-1} + \vartheta_i; \tag{140}$$

$$\eta_{i/i} = H_i x_i + \vartheta_i - H_i \hat{x}_{i/i} = H_i \delta_{i/i} + \vartheta_i. \tag{141}$$

Inequality (137), which is written in the following equivalent form:

$$J_i = J_0 - \gamma^{-2} \sum_{i=i_0+1}^{i_f} \left\| \eta_{i/i} \right\|_2^2 \geq 0, \tag{142}$$

where

$$J_0 = \sum_{i=i_0+1}^{i_f} \left(\|\xi_{i-1}\|_2^2 + \left\| \nu_{i/i-1} \right\|_2^2 \right), \tag{143}$$

in view of the constraint equation

$$\Phi_i x_{i-1} + \Gamma_i \xi_{i-1} - x_i = 0, \tag{144}$$

can be regarded as a quadratic estimation criterion for the vector x_i.

Optimization of estimates using criterion (142) should include minimization of the cost function (143) taking into account a priori observations and constraint (144)

$$\hat{x}_i = \arg\min_x J_0, \tag{145}$$

as well as ensuring positive definiteness of the quadratic form (142) based on a posteriori observations and the corresponding formation of the parameter γ.

A similar approach to optimizing estimates is considered in the work of Chernodarov [21]. This section presents a solution to problem (142) in expanded form, taking into account, as will be shown below, the adaptation of estimates to residuals (139).

It should be noted that the cost function (143) has a structure that allows the use of the mathematical apparatus of weighted least squares for optimization. This method makes it possible to allow for the "a priori" uncertainty of the residuals $\nu_{i/i-1}$ and disturbances ξ_i with the help of the matrices R_i and Q_i of weight coefficients, which normalize the relevant quadratic forms in relation (143), that is,

$$\|\nu_{i/i-1}\|^2_{R_i^{-1}} = \nu^{\mathrm{T}}_{i/i-1} R_i^{-1} \nu_{i/i-1}; \; \|\xi_{i-1}\|^2_{Q_{i-1}^{-1}} = \xi^{\mathrm{T}}_{i-1} Q_i^{-1} \xi_{i-1}. \tag{146}$$

We will consider estimates that minimize the penalty function (145) taking into account (146) and constraint (144) using the example of the traditional EKF [3] algorithm

$$\hat{x}_{i/i} = \hat{x}_{i/i-1} + P_{i/i} H_i^{\mathrm{T}} R_i^{-1} \nu_{i/i-1}, \tag{147}$$

where

$$P_{i/i} = P_{i/i-1} - P_{i/i-1} H_i^{\mathrm{T}} \left(H_i P_{i/i-1} H_i^{\mathrm{T}} + R_i\right)^{-1} H_i P_{i/i-1}; \tag{148}$$

Parameters $\hat{x}_{i/i-1}$ and $P_{i/i-1}$ are determined by Eqs. (73) and (74).

For realizing criterion (145), having regard to relations (147) and (148), the synchronization of the residuals $\nu_{i/i-1}$ and $\eta_{i/i}$ with respect to the appropriate estimate, for example, the estimate $\hat{x}_{i/i-1}$ becomes a necessity. This can be performed on the basis of an identity that follows from relation (141) after estimate (147) has been substituted in it, namely:

$$\eta_{i/i} = Z_i - H_i \hat{x}_{i/i} = Z_i - H_i \hat{x}_{i/i-1} - H_i P_{i/i} H_i^{\mathrm{T}} R_i^{-1} \nu_{i/i-1} = \tilde{Z}_i - \tilde{H}_i \hat{x}_{i/i-1}, \tag{149}$$

where $\tilde{Z}_i = Z_i - H_i K_i Z_i$; $\tilde{H}_i = H_i (I - K_i H_i)$; $K_i = P_{i/i} H_i^{\mathrm{T}} R_i^{-1}$.

Cost function (142) and condition (144), which reflects the dynamics of variation of the vector x_i, can be united into a single Euler-Lagrange structure [9] that is as follows:

$$\tilde{J} = J + \lambda_i^{\mathrm{T}} (\Phi_i x_{i-1} + \Gamma_i \xi_{i-1} - x_i), \tag{150}$$

where λ_i is the vector of Lagrange multipliers.

Such a representation of the cost function permits one to solve optimization problem (145) using Pontryagin's method [8, 10]. In accordance with this method, we form the Hamiltonian

$$\mathrm{H} = 0.5\left[\|\nu_i\|^2_{R_i^{-1}} - \|\eta_i\|^2_{\gamma^{-2}} + \|\xi_{i-1}\|^2_{Q_{i-1}^{-1}}\right] + \lambda_i^{\mathrm{T}}(\Phi_i x_{i-1} + \Gamma_i \xi_{i-1}) \tag{151}$$

and solve canonical equations of the form (119) with boundary conditions (120), where.

ν_i, η_i are residuals that are synchronized with respect to the estimate $\hat{x}_{i/i-1}$;

$$\|\eta_i\|^2_{\gamma^{-2}} = [\eta_1 \cdots \eta_l]\begin{bmatrix}\gamma^{-2} & 0 & 0\\ 0 & \gamma^{-2} & 0\\ 0 & 0 & \gamma^{-2}\end{bmatrix}\begin{bmatrix}\eta_1\\ \vdots\\ \eta_l\end{bmatrix}. \tag{152}$$

By calculating partial derivatives in accordance with expressions (119) and performing algebraic transformations, we obtain an expanded notation of the canonical equations

$$\hat{x}_{i/i-1} = a = \Phi_i \hat{x}_{i-1/i-1} - \Gamma_i Q_{i-1}\Gamma_i^{T}\Phi_i^{-T}\lambda_{i-1}; \tag{153}$$

$$\lambda_i = b = \Phi_i^{-T}\lambda_{i-1} + H_i^{T}R_i^{-1}\nu_i - H_i^{T}\gamma^{-2}\eta_i; \tag{154}$$

$$\hat{\xi}_{i-1} = -Q_{i-1}\Gamma_i^{T}\Phi_i^{-T}\lambda_{i-1}. \tag{155}$$

By analogy with the technique of solving TPBP (118)–(125), we define the partial derivatives included in the generalized Eqs. (95) and (96) for the implementation of the invariant embedding method

$$\frac{\partial a}{\partial c} = -\Phi_i P_{i-1/i-1} - \Gamma_i Q_{i-1}\Gamma_i^{T}\Phi_i^{-T}; \tag{156}$$

$$\frac{\partial b}{\partial c} = \Phi_i^{-\mathrm{T}} + \left(H_i^{\mathrm{T}}R_i^{-1}H_i - \tilde{H}_i^{\mathrm{T}}\tilde{\gamma}^{-2}\tilde{H}_i\right)\left(\Phi_i P_{i-1/i-1} + \Gamma_i Q_{i-1}\Gamma_i^{\mathrm{T}}\Phi_i^{-\mathrm{T}}\right). \tag{157}$$

After substituting relations (156) and (157) into Eqs. (95) and (96) and performing algebraic transformations taking into account $\hat{x}_{i/i}$:=$\hat{\Lambda}_{i/i}$; $P_{i/i}$:=$S_{i/i}$ and using the lemma on matrix inversion [13], we obtain the following error estimation algorithm.

Prediction: (73) and (74).

Updating:

$$\nu_{i/i-1} = Z_i - H_i \hat{x}_{i/i-1}; \tag{158}$$

$$\tilde{P}_{i/i} = \left(P_{i/i-1}^{-1} + H_i^{T}R_i^{-1}H_i\right)^{-1} = P_{i/i-1} - P_{i/i-1}H_i^{T}\left(H_i P_{i/i-1}H_i^{T} + R_i\right)^{-1}H_i P_{i/i-1}; \tag{159}$$

$$K_i = \tilde{P}_{i/i}H_i^{T}R_i^{-1}; \tag{160}$$

$$\hat{\hat{x}}_{i/i} = \hat{x}_{i/i-1} + K_i \nu_{i/i-1}; \tag{161}$$

$$P_{i/i} = \left(\tilde{P}_{i/i}^{-1} - \tilde{H}_i^{\mathrm{T}}\tilde{\gamma}^{-2}\tilde{H}_i\right)^{-1} = \tilde{P}_{i/i} + \tilde{P}_{i/i}\tilde{H}_i^{\mathrm{T}}\left(\tilde{\gamma}^2 - \tilde{H}_i\tilde{P}_{i/i}\tilde{H}_i^{\mathrm{T}}\right)^{-1}\tilde{H}_i\tilde{P}_{i/i}; \tag{162}$$

$$\tilde{K}_i = P_{i/i}\tilde{H}_i^{\mathrm{T}}\left(\tilde{\gamma}^2 - \tilde{H}_i P_{i/i}\tilde{H}_i^{\mathrm{T}}\right)^{-1}; \tag{163}$$

$$\eta_{i/i} = Z_i - H_i \hat{\tilde{x}}_{i/i}; \tag{164}$$

$$\hat{x}_{i/i} = \hat{\tilde{x}}_{i/i} - \tilde{K}_i \eta_i. \tag{165}$$

One can see that Eqs. (158)–(162) and (164) are equivalent to an EKF. Eqs. (162)–(165) to describe the functioning of a module intended for the protection of the integrity of the filter.

The proposed procedure for forming the γ^2 parameter, which guarantees a reliable estimation of the DS state, is based on the technology of iterative solution of Eq. (162), taking into account the necessary conditions for stable filtering (115), (116). As a generalized parameter that characterizes the adequacy of actual estimation errors and estimation errors predicted by H_∞ module (162)–(165), we can take the square of the normalized residual $\beta_j^2 = \eta_j^2/\alpha_j^2$, where $\alpha_j^2 = E\left[\eta_j^2\right] = H_j P_{i/i(j)} H_j^{\mathrm{T}} + R_j$; $\eta_i = \left[\eta_1 \cdots \eta_j \cdots \eta_l\right]^{\mathrm{T}}$; $H_i = \left[H_1^{\mathrm{T}} \ldots H_j^{\mathrm{T}} \ldots H_l^{\mathrm{T}}\right]^{\mathrm{T}}$; $P_{j(i/i)}$ is the value of the covariance matrix $P_{i/i}$ in relation (162) obtained after processing the jth component η_j of the vector η_i of residuals.

In practice, the divergence (13) of estimates and also the violation of condition (116) are connected with a decrease in the filter gain factor due to the degeneracy of the matrix $P_{j(i/i)}$. Because of this, it is advisable to increase the checked element of the specified matrix by such a value ΔQ_j that condition (116) turns into an identity, namely:

$$\frac{\eta_j^2}{H_j P_{i/i(j)} H_j^{\mathrm{T}} + R_j} = \frac{\eta_j^2}{H_j\left(P_{i/i(j-1)} + \Delta Q_j\right) H_j^{\mathrm{T}} + R_j} = \gamma_1^2. \tag{166}$$

It can be shown that the matrix elements ΔQ_j, which stabilize the solution of Eq. (166) with respect to the tolerance γ_1^2 when processing the jth observation, have the form:

$$\Delta Q_j = P_{i/i(j-1)} \tilde{H}_j^{T} q_j \tilde{H}_j P_{i/i(j-1)}, \tag{167}$$

where $q_j = d_j \alpha_j^2 / \left(\gamma_1^2 \tilde{\alpha}_j^2\right)$; $\tilde{\alpha}_j^2 = H_j P_{i/i(j-1)} \tilde{H}_j^{\mathrm{T}}$; $d_j = \hat{\beta}_j^2 - \gamma_1^2$.

It can be shown that the tunable parameter γ_j^2 in the $n \times n$ matrix $P_{i/i(j)}$, corresponding to the jth iteration of the solution of Eq. (162), has the form

$$\gamma_j^2 = \left[\left(\tilde{\alpha}_j^2\right)^2 \gamma_1^2 + d_j \alpha_j^2 \overline{\alpha}_j^2\right] / \left(d_j \alpha_j^2\right); j = \overline{1, n}. \tag{168}$$

Parameter (168) ensures the positive definiteness of the $P_{i/i(j)}$ matrix and the fulfillment of the control condition (116). If condition (116) is fulfilled, then $\Delta Q_j = 0$ and the value $P_{i/i(j)}$ of the covariance matrix remains unchanged.

5. Conclusions

In this chapter, the problem of joint estimation and control of parameters in observable dynamical systems (DS) is considered. Such problems arise in DS, the functioning of which is based on solving nonlinear differential equations for parameters and linearized equations for their errors. The presence of linearization errors and random disturbances in the DS and observations of parameters necessitates the use of an extended Kalman filter (EKF) in the estimation loop. The problem of using the EKF is related to ensuring the reliability of the estimation. To solve this problem, approaches based on non-quadratic criteria for optimizing observations are proposed. In contrast to traditional EKF, the use of such criteria allows one to form gain factors that depend on the residual of observations. The adaptive properties of the filters formed in this way ensure the parrying of parametric and stochastic uncertainties of the DS models.

Author details

Alexander V. Chernodarov
MAI University, Moscow, Russia

*Address all correspondence to: chernod@mail.ru

References

[1] Lawson G, Hanson R. Solving Least Squares Problems. New Jersey: Prentice – Hall; 1974

[2] Kwakernaak H, Sivan R. Linear Optimal Control Systems. New York: Wiley; 1972

[3] Maybeck P. Stochastic Models, Estimation and Control. Vol. 2. New York: Academic Press; 1982

[4] Fitzgerald RJ. Divergence of the Kalman filter. IEEE Transactions on Automatic Control. 1971;**16**(6):736-747

[5] Skorokhod A, Hoppensteadt F, Salehi H. Random Perturbation Methods with Applications in Science and Engineering. New York: Springer – Verlag; 2002

[6] Chernodarov A. Identification and an inverse filtering problem. In: 18th IFAC Symposium on System Identification (SYSID 2018), July 9–11, 2018, Stockholm, Sweden. IFAC-PapersOnLine. Amsterdam: Elsevier Publishing; 2018. pp. 66-71

[7] Detchmendy D, Sridhar R. Sequential estimation of states and parameters in noisy nonlinear dynamical systems. Transactions of the ASME Journal of Basic Engineering. 1966;**88D**:362-368

[8] Abutaleb A, Papaiouannou M. New results in Sridhar filtering theory: The discrete case. Journal of Optimization Theory and Applications. 1990;**64**(1): 5-14

[9] Speedy C, Brown R, Goodwin G. Control Theory: Identification and Optimal Control. Edinburgh: Oliver and Boyd; 1970

[10] Sage A, Melsa J. System Identification. New York: Academic Press; 1971

[11] Aoki M. Introduction to Optimization Techniques. New York: Collier – Macmillan; 1971

[12] Bierman G. Factorization Methods for Discrete Sequential Estimation. New York: Academic Press; 1977

[13] Sage A, White C. Optimum Systems Control. New Jersey: Prentice – Hall; 1977

[14] Chin L. Advances in adaptive filtering. In: Control and Dynamic Systems. New York: Academic Press; 1979. pp. 278-356

[15] Chernodarov A. State estimation of nonlinear dynamical systems on the basis of nonquadratic cost functions. In: 5th IFAC Symposium on Nonlinear Control Systems (NOLCOS'01), July 4–6, 2001, Saint-Petersburg, Russia, IFAC Proceedings. Vol . 3. Saint-Petersburg: Elsevier Publishing; 2002. pp. 905-909

[16] Chernodarov A. Non-quadratic optimization and adaptive robust filtering. In: 17th IFAC Workshop on Control Applications of Optimization (CAO 2018), Yekaterinburg, Russia, 15–19, October 2018, IFAC-PapersOnLine. Vol. 51. No. 32. Amsterdam: Elsevier Publishing; 2018. pp. 556-561

[17] Gertler J. Fault Detection and Diagnosis in Engineering Systems. New York: Marcel Dekker; 1998

[18] Chernodarov A. Application of the Kalman filter in monitoring, diagnosis, and fault parrying problems for observable dynamical systems. In: Applications and Optimizations of Kalman Filter and their Variants. London: IntechOpen; 2024. pp. 171-190

[19] Korn G, Korn T. Mathematical Handbook. New York: Mc Graw – Hill; 1968

[20] Souza C, Xie L. Robust H_∞ filtering. In: Control and Dynamic Systems. New York: Academic Press; 1994. pp. 323-377

[21] Chernodarov A. An H∞ technology for control of the integrity of the Kalman type of estimating filters with the use of adaptive robust procedures. In: 1st IFAC Conference on Modeling, Identification and Control of Nonlinear Systems (MICNON 2015), Saint Petersburg, Russia, June 24–26, 2015. IFAC PapersOnline. Vol. 48. No. 11. New York: Curran Associates, Inc.; 2015. pp. 348-353

Chapter 2

Parametric Study of the Energy Dissipation by a Suspension System

Uchenna Diala and Godwin Sani

Abstract

This study examines the energy dissipation characteristics of a nonlinear vehicle suspension system. The suspension system is described as a nonlinear mass-spring-damper system, which is developed using the output frequency response function (OFRF) technique. This approach allows for the establishment of a direct relationship between the output spectrum of the system (or performance metric) and the nonlinear parameters of interest. The OFRF polynomial, related to the level of energy dissipation, was formulated in terms of the relevant nonlinear parameters. The findings indicate that the OFRF model for energy dissipation can be utilised to estimate the heat generated for specific design parameters. This could offer valuable insights for designing vibration-based energy harvesters, as some of the energy lost as heat could potentially be converted into electrical energy.

Keywords: vibration isolation system, nonlinear suspension system, OFRF, nonlinear systems, energy dissipation, nonlinear damping

1. Introduction

In response to the increasing demands of contemporary society, new transportation methods are being created to enhance ride quality, safety, and comfort. These advancements are also anticipated to improve efficiency and guarantee cost-effectiveness across all transportation modes. Various ground vehicles encounter different degrees of vibration due to the unevenness of the road surface. These vibrations travel to the vehicle's body (including the occupants) through its tyres and suspension systems (shock absorbers). Such vibrations can greatly diminish ride comfort and quality, and even affect ride stability [1–4]. The research in [2] examined the regulation of roll and vertical vibrations in a seat suspension system (a two-layer multi-DOF active seat suspension) caused by uneven road surfaces beneath both tyres. Two types of controllers, a non-singular terminal sliding mode controller and an H_∞ controller, were implemented. Simulation and experimental studies validated the effectiveness of both the seat suspension system and the employed control methods. In [3], the authors explored how asymmetrical damping and geometric nonlinearities in a vehicle model's suspension system influence ride comfort. They demonstrated, through both numerical and experimental means, that combining asymmetric and geometric nonlinearities enhanced passengers' ride comfort and the performance of the suspension system. In [4], a vibration absorber integrated into the wheel, consisting of a spring, an annular rubber bushing, and an adjustable damper, was proposed to mitigate vibrations produced by the motor, using an improved particle swarm optimisation (IPSO) algorithm to determine its parameters. Two control strategies, a linear quadratic regulator (LQR) and a fuzzy proportional-integral-derivative (PID) approach, were utilised to manage the suspension damper and the in-wheel damper, respectively, both showing excellent performance. Vehicle suspension systems are designed to curb the effect of adverse ground vibrations and typically comprise a spring, damper and set of linkages [5]. However, the mass is made up of the vehicle body and passengers [5]. A key property of an isolation spring lies in its capacity to absorb the vibration energy generated from ground surface irregularities, and store the energy as potential energy. Nonetheless, springs made from rubber, coil, and leaf springs are not very effective at dissipating the energy absorbed. This limitation can lead to undesirable performance of the spring before it eventually gets to its breaking point, which highlights the necessity for a well-designed damping structure [6]. A damping structure is crucial in isolating vibrations from the vehicle's mass, as it attenuates vertical motions by converting the transmitted vibration energy into heat energy [6, 7]. This action ensures that the lifespan of the stiffness element is prolonged and the structural integrity of the system of interest is preserved. Designing an effective isolation system is vital in suspension systems, to provide a satisfactory ride comfort [7–9]. Lu and his colleagues, in [10], categorised nonlinear dampers into three types: nonlinear stiffness dampers, nonlinear-stiffness nonlinear-damping dampers, and nonlinear damping dampers. Drawing upon this classification, the authors examined three dampers frequently utilised in practical engineering - nonlinear energy sink (NES), particle impact damper (PID), and nonlinear viscous damper (NVD), respectively [10]. **Figure 1** shows the application of the suspension system.

Due to the intrinsic nonlinearity inherent in structural-based isolation systems, there has been an increased focus on nonlinear vibration isolation systems. As a result,

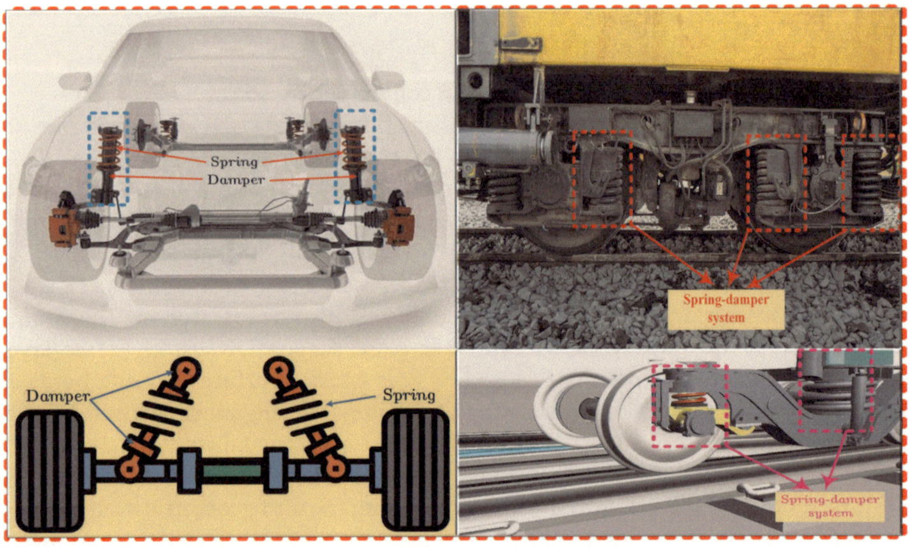

Figure 1.
Applications of the suspension system.

various designs of nonlinear isolators of engine mounts have been extensively documented in the literature [11–15].

The authors in [11] analysed two models of ride comfort for a multibody dynamic system, differentiated by their damping properties – one with nonlinear damping and the other with its corresponding linear damping. The responses measured from the driver's seat were juxtaposed using these models, revealing that the nonlinear damping model yielded more precise results compared to the equivalent damping model. Conversely, in [12], the authors introduced an innovative nonlinear quasi-zero-stiffness (QZS) vibration isolator, wherein the suspension system exhibited nonlinearity due to the arrangement of inclined springs. The QZS isolator is defined by its high-static and low-dynamic stiffness. Findings indicated that the QZS vibration isolator, when equipped with an active controller, provided excellent isolation performance in response to large amplitude excitations. In [13], a computational analysis was performed on a QZS isolator model akin to that in [12]. Their research examined various road profiles as referenced in ISO 8608. The results demonstrated that the proposed system successfully diminished external excitations by approximately 92% and 88% on average for road classes B-D and E-H, respectively. The effect of vehicle velocity, along with a nonlinear suspension system, on the dynamic behaviour of a Bernoulli-Euler bridge, was explored in [14]. It was determined that the speed of the vehicle, as well as the stiffness and damping behaviour of its suspension, are the primary factors that significantly affect the dynamic behaviour of a bridge, under force excitations. In [15], the effects of asymmetrical damping and geometric nonlinearities on a vehicle's suspension system performance were investigated. Both numerical and experimental findings confirmed that geometric nonlinearity has a considerable influence on the behaviour of damping and stiffness characteristics in a suspension system, thereby enhancing its capacity to reduce vibrations.

There are various analytical techniques available for examining nonlinear suspension systems, including averaging method [16, 17], describing function method [18, 19], and harmonic balance method [20, 21]. Nonetheless, the above-mentioned techniques fail to clearly illustrate the connection between performance metrics of the suspension system and the nonlinear parameters of the system. The frequency-based approach known as the Output Frequency Response Function (OFRF) [22–25] is a frequency-based technique, which allows for the formulation of the output spectrum of the system of interest, as a function of the system's nonlinear parameters of interest. This approach is utilised in this research, since it supports a frequency-based analysis and design of nonlinear dynamical systems.

A passive vibration isolation model, described as a single degree of system (SDOF), and featuring both cubic stiffness and damping properties was examined in [26]. The averaging perturbation technique was utilised to investigate how nonlinearities affect the dynamic behaviour of the system. Similarly, Peng and Lang analysed the same nonlinear passive mount in [25], employing the OFRF approach for a theoretical assessment of the output frequency response behaviour concerning the system's nonlinearities, which was subsequently confirmed through numerical analysis. A related system was considered in [27], wherein the impact of the nonlinear parameters of the system on the displacement transmissibilities (absolute and relative) were examined.

In this research, the vehicle suspension system depicted in **Figure 2**, which closely resembles those discussed in articles [25–27], was examined. Nonetheless, analytical study and design of the suspension system were performed using the OFRF approach. Ideal design parameter values were derived to meet the specified design criteria, and these designed parameters are corroborated through numerical simulations.

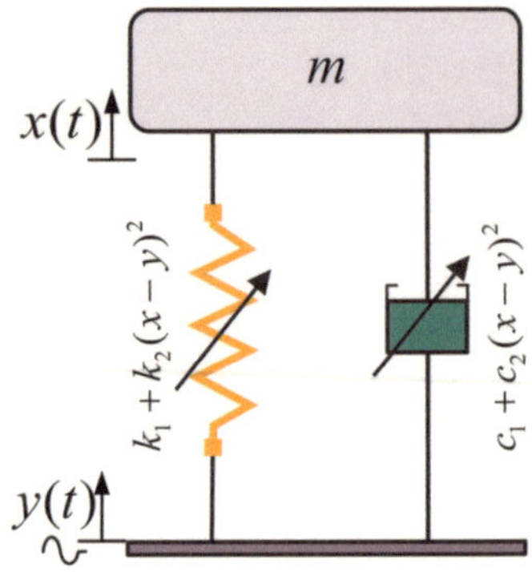

Figure 2.
A base-excited nonlinear suspension system.

Additionally, in this study, the energy dissipation performance of a vehicle suspension system was assessed using the OFRF technique, for the first time. The influence of the nonlinear behaviour of the suspension system, on the energy dissipated by the isolation system, was investigated. Following this, the investigation uses the OFRF method, where the nonlinear system parameters are identified to achieve a specified energy dissipation level within the suspension system.

The rest of the chapter is organised as follows: Section 2 provides the model formulation of the suspension system to be investigated, while the OFRF method is introduced in Section 3. Section 4 explains the deduction of the OFRF depiction for the output response of the system, whereas Section 5 illustrates the design of the system of interest and the system optimisation process, based on the requirement. The effects of the system parameters are demonstrated in Section 6, using numerical simulation, while the analytical study of the energy dissipation level, for the developed suspension system, as well as the effects of system nonlinearities on the dissipation level, are considered in Section 7. Finally, conclusions are reached in Section 8.

2. System design

Figure 2 illustrates a schematic of the nonlinear Single Degree of Freedom (SDOF) suspension system examined in this study. This system features a suspended oscillating sprung mass, denoted as m, which represents the quarter mass of a vehicle body, with an absolute displacement of $x(t)$ resulting from a base motion of $y(t)$. The suspension system consists of a nonlinear damping system with damping coefficients, c_1, and c_2 and nonlinear stiffness, with coefficients k_1 and k_2.

Eq. (1) provides the equation of motion for the suspension system as

$$m\ddot{z} + c\dot{z} + kz = -m\ddot{y} \tag{1}$$

where $z = x - y$ is the relative displacement between the base and the sprung mass, while c and k are the damping and stiffness functions, respectively. The damping and stiffness functions are described as

$$\begin{cases} k = k_1 + k_2 z^2 \\ c = c_1 + c_2 z^2 \end{cases} \tag{2}$$

Given that the base excitation is harmonic, that is

$$y = Y\sin(\omega t) \tag{3}$$

then Eq. (1) becomes

$$\begin{cases} m\ddot{z} + (c_1 + c_2 z^2)\dot{z} + (k_1 + k_2 z^2)z = m\omega^2 Y\sin(\omega t) \\ f_{\text{trans}} = (c_1 + c_2 z^2)\dot{z} + (k_1 + k_2 z^2)z \end{cases} \tag{4}$$

where ω and Y are the excitation frequency and amplitude of the base displacement, respectively, while f_{trans} is the force transmitted through the vibration isolation system (a parallel combination of the spring and damper elements).

To enhance ride comfort, it is essential to minimise the transmitted force. The output spectrum of the transmitted force reaches its peak at the resonant frequency ω_r, which is the critical frequency of interest. The objective will be to minimise the transmitted force f_{trans}, at this resonant frequency. We will employ the procedure outlined by Zhu and Lang in [28] for the analysis and design of a Single Input Multiple Output (SIMO) system using the OFRF method.

In the following section, we will introduce the OFRF method while considering a general Volterra system, which is described by a nonlinear differential equation.

3. The output frequency response function (OFRF)

The system in Eq. (5) is used to describe a class of Volterra systems

$$\sum_{\bar{q}=1}^{Q}\sum_{v=0}^{\bar{q}}\sum_{r_1,\dots r_{\bar{q}}=0}^{R} c_{v,\bar{q}-v}(r_1,\cdots r_{\bar{q}})\prod_{i=1}^{v}\frac{d^{r_i}y(t)}{dt^{r_i}}\prod_{i=v+1}^{\bar{q}}\frac{d^{r_i}u(t)}{dt^{r_i}} = 0 \tag{5}$$

where, Q is the maximum degree of nonlinearity, with respect to the input $u(t)$ and output $y(t)$ parameters of the system, and R is the order of the derivative. In accordance with the OFRF methodology cited in Refs. [22–25], the output frequency response of the system delineated in Eq. (5) can be characterised by a polynomial function. This function is expressed in relation to the nonlinear parameters inherent to the system, described as

$$Y(\mathrm{j}\omega) = \sum_{\gamma_1=0}^{z_1}\cdots\sum_{\gamma_{S_N}=0}^{z_{S_N}} \Upsilon_{(\gamma_1,\dots,\gamma_{S_N})}(\mathrm{j}\omega)\eta_1^{\gamma_1}\dots\eta_{S_N}^{\gamma_{S_N}} \tag{6}$$

where, z_i are the maximum order of η_i, for $i = 1, \dots, S_N$, in the output spectrum $Y(\mathrm{j}\omega)$, and $\Upsilon_{(\gamma_1,\dots,\gamma_{S_N})}(\mathrm{j}\omega)$ are frequency functions, which are also referred to as the OFRF coefficients. These functions are complex values are dependent on the system's input and linear characteristic parameters, where $\gamma_i = 0, \dots, z_i$ and $i = 1, \dots, S_N$. Furthermore, $\eta_1^{\gamma_1}\dots\eta_{S_N}^{\gamma_{S_N}}$ is a set of monomials also known as the OFRF structure. They are derived in terms of the nonlinear characteristic parameters of the system. It should be noted that $\Upsilon_{(\gamma_1,\dots,\gamma_{S_N})}(\mathrm{j}\omega)$ also has the coefficients $\eta_1^{\gamma_1}\dots\eta_{S_N}^{\gamma_{S_N}}$ with the same dimension, and needs to be evaluated in order to obtain the OFRF representation of system (5).

To begin with, the monomials must be identified by employing a recursive algorithm presented by Peng et al. [25]. For a Single-Input-Multiple-Output (SIMO) system, Zhu and Lang have outlined a detailed process in [28]. Let the collection of monomials in the OFRF representation of the nth-order output spectrum be represented as $\mathfrak{M}$ and let the frequency function vector be represented as $\Phi(j\omega)$, then the OFRF can be defined as

$$Y(j\omega) = \mathfrak{M} \cdot \Phi(j\omega)^{T} \tag{7}$$

where

$$\mathfrak{M} = \bigcup_{n=1}^{N} M_n \tag{8}$$

Here, the monomials, M_n can be determined using [25].

$$\begin{aligned} M_n = & \left[\bigcup_{r_1, \dots, r_n=0}^{R} \left[c_{0,n}(r_1, \dots, r_n) \right] \right] \\ & \cup \left[\bigcup_{\bar{q}-v=1}^{n-1} \bigcup_{v=1}^{n-(\bar{q}-v)} \bigcup_{r_1, \dots, r_n=0}^{R} \left(\left[c_{v,(\bar{q}-v)}(r_1, \dots, r_{\bar{q}}) \right] \otimes M_{n-(\bar{q}-v),v} \right) \right] \\ & \cup \left[\bigcup_{v=2}^{n} \bigcup_{r_1, \dots, r_n=0}^{R} \left(\left[c_{v,0}(r_1, \dots, r_{\bar{q}}) \right] \otimes M_{n,v} \right) \right] \end{aligned} \tag{9}$$

where

$$M_{n,v} = \bigcup_{i=1}^{n-v+1} (M_i \otimes M_{n-i,v-1}), M_{n,1} = M_n, M_1 = 1 \tag{10}$$

The symbol '$\otimes$' is the Kronecker product. Therefore, the class of coefficients of the OFRF model can be determined as $\mathfrak{M} = \bigcup_{n=1}^{N} M_n$.

In Section 4, the OFRF approach is applied to establish a relationship between the nonlinear system parameters and the output performance metrics of system (4). Following this, an analysis and design of the system is conducted to determine the best system parameters that provide the best performance, according to the design requirements. Additionally, the impact of the nonlinear parameters, c_2 and k_2, on the performance metrics of the system will be examined through numerical simulations.

4. Formulation of the OFRF polynomial

The OFRF output response of system (4) is derived in relation to the key nonlinear parameters, c_2 and k_2. As shown in Eq. (6), the OFRF of the nonlinear system (5) is expressed as a polynomial function of the nonlinear characteristic parameters of the system. The coefficients of the OFRF are dependent on the output frequency of the system, which is determined by the specific form of Eq. (5), the input spectrum of the system, and the linear characteristic parameters. To utilise the OFRF representation of system (4) for both analysis and design, it is essential to establish its coefficients and monomials initially. The analyses below have been conducted using the system parameters detailed in **Table 1**.

Parameter	Value	Unit
m	0.2	kg
c_1	1.5	$\mathrm{N\,s\,m^{-1}}$
Y	0.05	m

Table 1.
System parameters.

It can be inferred that the system in Eq. (4), is the instance of system (5), with $c_{0,1}(0) = -m\omega^2 Y$, $c_{1,0}(2) = m$, $c_{1,0}(0) = k_1$, $c_{1,0}(1) = c_1$, $c_{3,0}(0,0,1) = c_2$, $c_{3,0}(0,0,0) = k_2$, else $c_{p,q}(\cdot) = 0$.

4.1 Ascertaining the OFRF structure

In [25, 28], the algorithm presented for finding the OFRF representation, for single-input-multiple-output systems, is used. The set of monomials, $\mathfrak{M}$, up to N = 7th order, obtained using the recursive algorithm therein, is applied to system (4). This yields

$$\mathfrak{M} = \bigcup_{n=1}^{7} \mathrm{M_n} = \left[1\ c_2\ k_2\ c_2^2\ c_2k_2\ k_2^2\ c_2^3\ c_2^2k_2\ c_2k_2^2\ k_2^3\right] \tag{11}$$

It is important to highlight that both outputs of system (4) yield the same set of monomials. This suggests that the OFRF representations for the output responses of system (4) can be represented as

$$\begin{cases} Z(\mathrm{j}\omega) = \mathfrak{M} \cdot \underline{\Phi}(\mathrm{j}\omega) \\ F_{\mathrm{trans}}(\mathrm{j}\omega) = \mathfrak{M} \cdot \underset{\sim}{\Phi}(\mathrm{j}\omega) \end{cases} \tag{12}$$

where $\underline{\Phi}(\mathrm{j}\omega)$ and $\underset{\sim}{\Phi}(\mathrm{j}\omega)$ represent the OFRF coefficients of the output responses of system (4), $Z(\mathrm{j}\omega)$and $F_{\mathrm{trans}}(\mathrm{j}\omega)$, respectively. The next task is to determine the coefficients of the OFRF polynomial.

4.2 Deduction of the coefficients of the OFRF polynomial

To deduce the OFRF coefficients for the system responses, $Z(\mathrm{j}\omega)$ and $F_{\mathrm{trans}}(\mathrm{j}\omega)$, the procedure provided in the Refs. [22–25, 28] are followed. The least-square method is used to extract the coefficients $\underline{\Phi}(\mathrm{j}\omega)$ and $\underset{\sim}{\Phi}(\mathrm{j}\omega)$ as

$$\begin{cases} \underline{\Phi}(\mathrm{j}\omega)^{\mathrm{T}} = \left(\psi^{\mathrm{T}}\psi\right)^{-1}\psi^{\mathrm{T}} \cdot \tilde{Z}(\mathrm{j}\omega) \\ \underset{\sim}{\Phi}(\mathrm{j}\omega)^{\mathrm{T}} = \left(\psi^{\mathrm{T}}\psi\right)^{-1}\psi^{\mathrm{T}} \cdot \tilde{F}_{\mathrm{trans}}(\mathrm{j}\omega) \end{cases} \tag{13}$$

where,

$$\underline{\Phi}(\mathrm{j}\omega)^{\mathrm{T}} = \begin{bmatrix} \underline{\phi}_{0,0}(\mathrm{j}\omega) \\ \underline{\phi}_{1,0}(\mathrm{j}\omega) \\ \vdots \\ \underline{\phi}_{0,3}(\mathrm{j}\omega) \end{bmatrix}, \quad \underset{\sim}{\Phi}(\mathrm{j}\omega)^{\mathrm{T}} = \begin{bmatrix} \underset{\sim 0,0}{\phi}(\mathrm{j}\omega) \\ \underset{\sim 1,0}{\phi}(\mathrm{j}\omega) \\ \vdots \\ \underset{\sim 0,3}{\phi}(\mathrm{j}\omega) \end{bmatrix},$$

$$\tilde{Z}(\mathrm{j}\omega) = \begin{bmatrix} Z(\mathrm{j}\omega)_{(1)} \\ Z(\mathrm{j}\omega)_{(2)} \\ \vdots \\ Z(\mathrm{j}\omega)_{(Q)} \end{bmatrix}, \quad \tilde{F}_{\mathrm{trans}}(\mathrm{j}\omega) = \begin{bmatrix} F_{\mathrm{trans}}(\mathrm{j}\omega)_{(1)} \\ F_{\mathrm{trans}}(\mathrm{j}\omega)_{(2)} \\ \vdots \\ F_{\mathrm{trans}}(\mathrm{j}\omega)_{(Q)} \end{bmatrix} \tag{14}$$

and

$$\psi = \begin{bmatrix} 1 & c_{2(1)} & k_{2(1)} & c_{2(1)}^2 & c_{2(1)}k_{2(1)} & k_{2(1)}^2 & c_{2(1)}^3 & c_{2(1)}^2 k_{2(1)} & c_{2(1)}k_{2(1)}^2 & k_{2(1)}^3 \\ & & & & & \vdots & & & & \\ 1 & c_{2(Q)} & k_{2(Q)} & c_{2(Q)}^2 & c_{2(Q)}k_{2(Q)} & k_{2(Q)}^2 & k_{2(Q)}^2 & c_{2(Q)}^2 k_{2(Q)} & c_{2(Q)}k_{2(Q)}^2 & k_{2(Q)}^3 \end{bmatrix}$$

$Z(\mathrm{j}\omega)$ and $F_{\mathrm{trans}}(\mathrm{j}\omega)$ denote the output spectra of system (4), when $c_{2(i)} = c_2(i)$ and $k_{2(i)} = k_2(i)$, for $i = 1, 2, \cdots, Q$, where Q = 25 for a varied set of $c_2 \in [0, 28]$ N s m^{-3} and $k_2 \in [0, 2] \times 10^2$ N m^{-3}. Therefore, for N = 7, the output responses of system (4), in the OFRF form, is given as

$$\begin{cases} Z(\mathrm{j}\omega) = \sum\limits_{\overline{n}=0}^{3} \sum\limits_{\overline{m}=0}^{\overline{n}} \underline{\phi}_{\overline{m},\overline{n}-\overline{m}}(\mathrm{j}\omega) c_2^{\overline{m}} k_2^{\overline{n}-\overline{m}} \\ \quad = \underline{\phi}_{0,0}(\mathrm{j}\omega) + \underline{\phi}_{1,0}(\mathrm{j}\omega)c_2 + \underline{\phi}_{0,1}(\mathrm{j}\omega)k_2 \\ \quad + \underline{\phi}_{2,0}(\mathrm{j}\omega)c_2^2 + \underline{\phi}_{1,1}(\mathrm{j}\omega)c_2k_2 + \underline{\phi}_{0,2}(\mathrm{j}\omega)k_2^2 \\ \quad + \underline{\phi}_{3,0}(\mathrm{j}\omega)c_2^3 + \underline{\phi}_{2,1}(\mathrm{j}\omega)c_2^2k_2 + \underline{\phi}_{1,2}(\mathrm{j}\omega)c_2k_2^2 \\ \quad + \underline{\phi}_{0,3}(\mathrm{j}\omega)k_2^3 \\ \text{and} \\ F_{\mathrm{trans}}(\mathrm{j}\omega) = \sum\limits_{\overline{n}=0}^{3} \sum\limits_{\overline{m}=0}^{\overline{n}} \underset{\sim \overline{m},\overline{n}-\overline{m}}{\phi}(\mathrm{j}\omega) c_2^{\overline{m}} k_2^{\overline{n}-\overline{m}} \\ \quad = \underset{\sim 0,0}{\phi}(\mathrm{j}\omega) + \underset{\sim 1,0}{\phi}(\mathrm{j}\omega)c_2 + \underset{\sim 0,1}{\phi}(\mathrm{j}\omega)k_2 \\ \quad + \underset{\sim 2,0}{\phi}(\mathrm{j}\omega)c_2^2 + \underset{\sim 1,1}{\phi}(\mathrm{j}\omega)c_2k_2 + \underset{\sim 0,2}{\phi}(\mathrm{j}\omega)k_2^2 \\ \quad + \underset{\sim 3,0}{\phi}(\mathrm{j}\omega)c_2^3 + \underset{\sim 2,1}{\phi}(\mathrm{j}\omega)c_2^2k_2 + \underset{\sim 1,2}{\phi}(\mathrm{j}\omega)c_2k_2^2 \\ \quad + \underset{\sim 0,3}{\phi}(\mathrm{j}\omega)k_2^3 \end{cases} \tag{15}$$

Considering the output spectra related to the design parameters, as defined in Eq. (15), a parameter optimisation method can be performed efficiently. For instance, regarding the resonant frequency, given as $\omega_r = 50\,\text{rad.s}^{-1}$, the OFRF model of the system's output spectra at this frequency are established as

$$\begin{aligned}Z(j\omega_r) &= (-0.1585 - 0.2344i) + (0.0022 + 0.0040i)c_2\\ &+(2.6195\times10^{-4} - 1.6383\times10^{-4}i)k_2 + (-3.7058\times10^{-5} - 1.2109\times10^{-4}i)c_2^2\\ &+(-1.2745\times10^{-5} + 5.3623\times10^{-6}i)c_2k_2 + (7.1624\times10^{-8} + 2.6437\times10^{-7}i)k_2^2\\ &+(1.1426\times10^{-7} + 1.6910\times10^{-6}i)c_2^3 + (2.3307\times10^{-7} - 4.9742\times10^{-8}i)c_2^2k_2\\ &+(3.8650\times10^{-10} - 9.5687\times10^{-9}i)c_2k_2^2 + (-1.9915\times10^{-10} + 1.2264\times10^{-11}i)k_2^3\end{aligned} \tag{16}$$

and

$$\begin{aligned}F_{\text{trans}}(j\omega_r) &= (-6.1703\times10^{1} - 1.2908\times10^{2}i)\\ &+(1.1198 + 2.0110i)c_2 + (0.1309 - 0.0820i)k_2\\ &+(-0.0186 - 0.0606i)c_2^2 + (-0.0064 + 0.0027i)c_2k_2\\ &+(3.5977\times10^{-5} + 1.3211\times10^{-4}i)k_2^2 + (5.7607\times10^{-5} + 8.4602\times10^{-4}i)c_2^3\\ &+(1.1653\times10^{-4} - 2.4955\times10^{-5}i)c_2^2k_2 + (1.9106\times10^{-7} - 4.7826\times10^{-6}i)c_2k_2^2\\ &+(-9.9691\times10^{-8} + 6.3415\times10^{-9}i)k_2^3\end{aligned} \tag{17}$$

respcctively.

In this case, the magnitude of the output responses, $|Z(j\omega_r)|$and $|F_{\text{trans}}(j\omega_r)|$, of system (4), are plotted for a set of $c_2 \in [0, 28]$ N.s.m-3 while $k_2 = 240\text{N.m}^{-3}$is fixed. This is demonstrated in **Figure 3**.

In **Figure 3**, the results shown were determined by employing the OFRF models for the output responses, given in Eqs. (16) and (17), and are matched with those derived using simulation studies. It is important to note that some of these results were achieved with parameter values that exceeds those initially used to establish the OFRF representations. The findings demonstrate a strong correlation between the OFRF-based results and those from simulations.

This demonstrates the effectiveness of the OFRF approach, which provides a reliable representation of the actual system output response. Consequently, the derived OFRF models for the output responses of system (4), could be leveraged for both analytical and design purposes. By utilising the OFRF method, a clear association between the output responses of interest and the design parameters, can be demonstrated.

It is crucial to note that an incorrect OFRF model of the definite system output responses will yield an unsatisfactory analytical performance. This poor performance could arise from significant numerical errors in the deductions of the OFRF coefficients. This typically happens when very large parameters are selected, the parameters exhibit considerable numerical differences, or when the selected parameters span a broad range. Consequently, the matrix $\mathfrak{M}$ has very large elements. In addition, computational errors may be significant when the matrix inverse of $\mathfrak{M}$ is calculated. The steps described in [23, 28] can be followed to solve this problem.

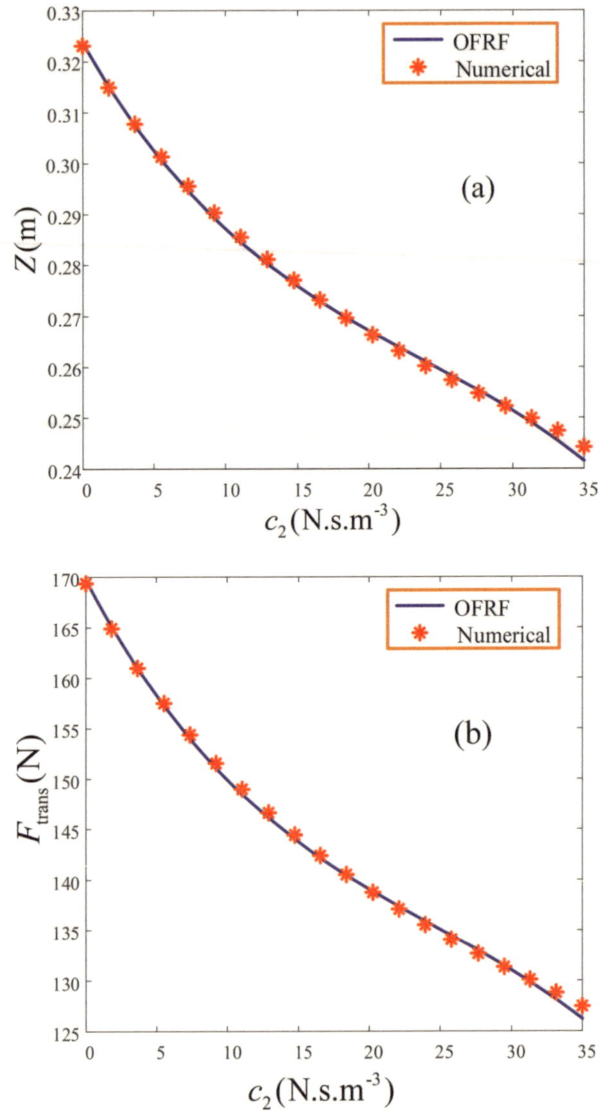

Figure 3.
Comparison between the OFRF and numerical simulation results, for (a) $Z(j\omega_r)$ *and (b)* $F_{trans}(j\omega_r)$*, output responses of system (4), respectively.*

5. System optimisation

The goal at this point is to reduce the undesired force that is transmitted through the isolation components (spring-damper system) at the resonant frequency, $|F_{trans}(j\omega_r)|$, to an acceptable level. Let the acceptable limit of the force propagated along the isolation system be defined as $|\mathbb{F}_{trans}(j\omega_r)|$; thus, the optimisation problem can be framed as

$$\min_{[c_2,k_2]} \ |F_{trans}(j\omega_r) - \mathbb{F}_{trans}(j\omega_r)|$$
$$\text{subject to} \ \begin{cases} c_2 - 35 \leq 0 \\ k_2 - 2.4 \times 10^2 \leq 0 \end{cases} \tag{18}$$

By employing the derived OFRF of Eq. (17), the association between the system parameters of interest, c_2 and k_2, and the output response, $F_{\text{trans}}(j\omega_r)$, is demonstrated and presented in **Figures 4** and **5**. Observing **Figure 5** shows that a contour map could facilitate the provision of a design methodology. The outcomes of two selected design specifications are shown in **Table 2**. To deduce the actual output response, $F_{\text{trans}}(j\omega_r)$, the parameters of interest are substituted into the system model (4) and, using MATLAB, is numerically simulated.

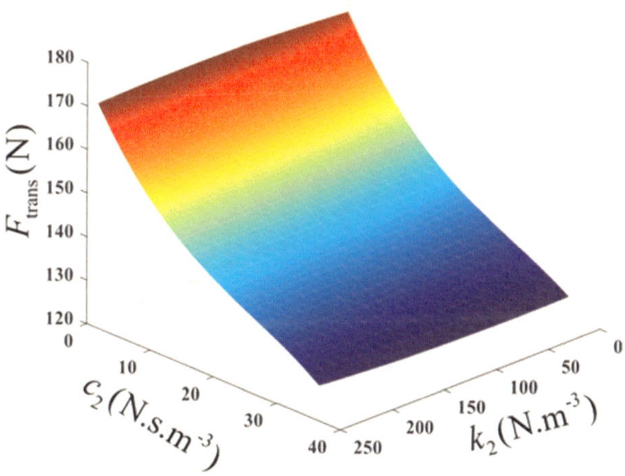

Figure 4.
Output response curve of the propagated force, $|F_{\text{trans}}(j\omega_r)|$.

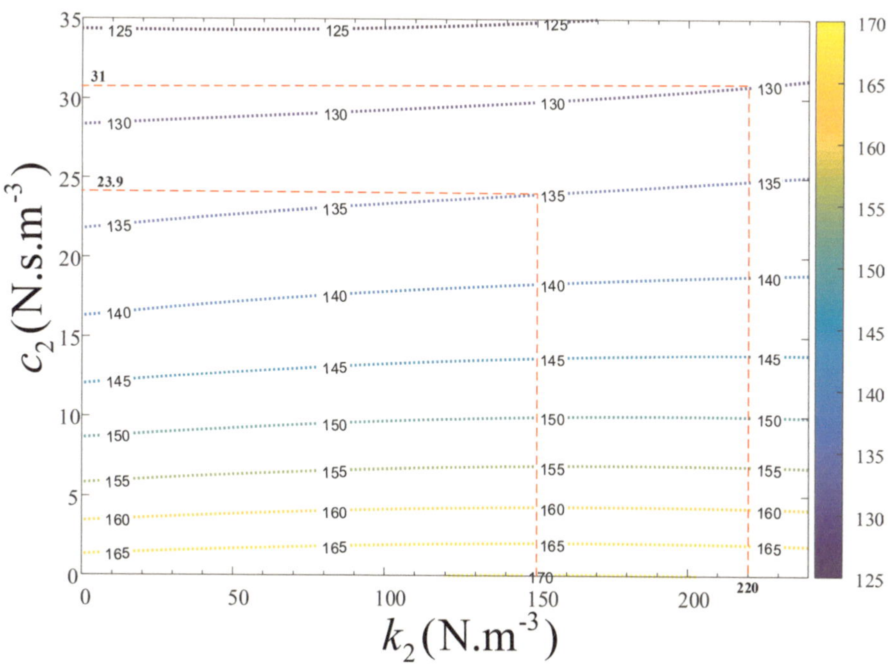

Figure 5.
A Contour map of the output response of the propagated force, $|F_{\text{trans}}(j\omega_r)|$.

Desired output response, $|\mathbb{F}_{\mathbf{trans}}(\mathbf{j}\omega_r)|$ (N)	Selected $k_2(\mathrm{N.m^{-3}})$	Corresponding $c_2(\mathrm{N.s.m^{-3}})$	Actual output response $|F_{\mathbf{trans}}(\mathbf{j}\omega_r)|$ (N)	Percentage error (%) $\left[\frac{|\mathbb{F}_{\mathrm{trans}}(\mathrm{j}\omega_r)|-|F_{\mathrm{trans}}(\mathrm{j}\omega_r)|}{|F_{\mathrm{trans}}(\mathrm{j}\omega_r)|}\right] \times 100$
135	150	23.9	134.8924	7.9767×10^{-2}
130	220	31	130.1583	-1.2162×10^{-1}

Table 2.
Results from the optimisation problem.

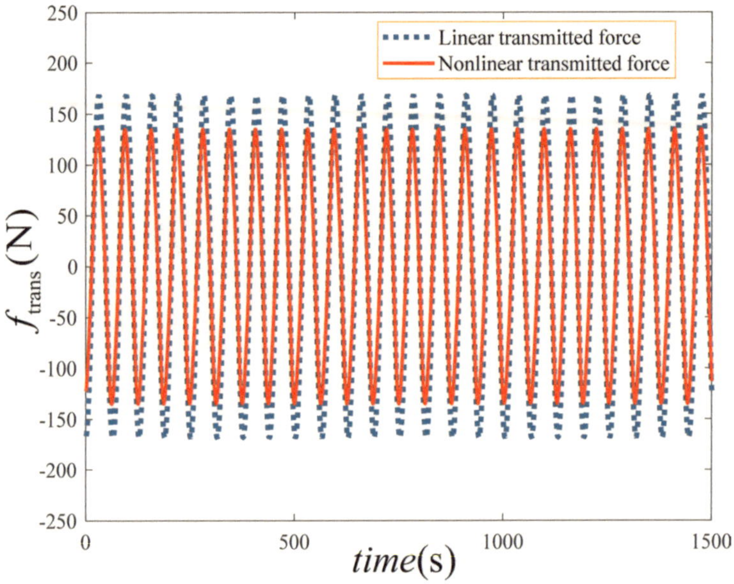

Figure 6.
System output responses, at resonance, under linear and nonlinear situations.

The second design specification was realised for a set of parameters outside the set used to determine the OFRF, i.e., $c_2 \in [0, 28]$N s m^{-3} and $k_2 \in [0, 2] \times 10^2$ N m^{-3}. It is observed in **Table 2**, that the actual output responses, determined by utilising the designed parameters, c_2 and k_2, obviously agrees with the specified output responses, with insignificant percentage errors. This confirms the capability of the OFRF-based optimisation method.

From **Figure 6**, the time history graph of the propagated force, at the resonant frequency, shows that the nonlinear vibration isolation system outperforms the linear equivalent, i.e., where the nonlinear parameters, $c_2 = k_2 = 0$. Furthermore, observing the time history closely, it is seen that the desired force, propagated at the resonant frequency, is realised. The suppression of the force propagated through the isolation system, to a suitable level, $|F_{\mathrm{trans}}(\mathrm{j}\omega_r)|$, will also result in the reduction of the system output response, $|Z(\mathrm{j}\omega_r)|$.

Additionally, while the system design has been executed at the resonant frequency, it is essential to confirm that the established nonlinear optimal parameters sustain similar performance on the system output spectrum throughout the entire frequency range. This is demonstrated in **Figure 7**, which compares the performance of the nonlinear design to its linear equivalent (i.e., when the conditions are linear) on the

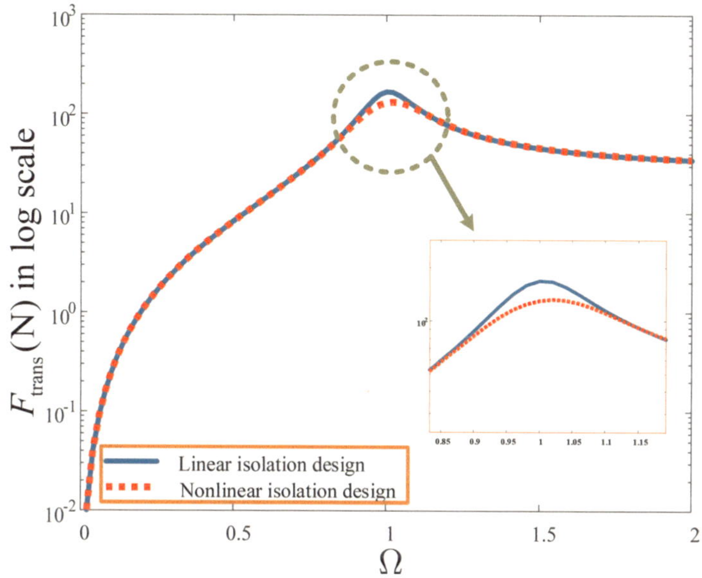

Figure 7.
System output response, under linear and nonlinear conditions.

system output spectrum. It is clear that the optimal parameters not only achieve the desired output spectrum at the resonant frequency but also effectively reduce vibration levels across other frequencies.

The OFRF method was used for the analysis and design of the system outlined above. This approach clearly enables a structured analysis and design of nonlinear Volterra systems compared to alternative methods. The following section presents the specific impact of each system parameter on the output spectra, as examined in this study. Each parameter is adjusted while keeping the others constant. These assessments were carried out through numerical simulations.

6. Simulation studies

To illustrate the impacts of c_1, c_2, and k_2 on the output response of system (4), numerical analyses were conducted, with the resulting findings displayed in **Figures 8–10**, respectively. **Figure 8** reveals the well-documented effect of the linear damping characteristic parameter on the output spectra of system (4), where effective isolation performance is observed within the resonant region. However, its negative impact in the area requiring isolation ($\Omega \gg 1$) at high damping levels indicates the necessity for incorporating a nonlinear damping component. In **Figure 9**, the influence of k_2 is noticeable in the resonant region. It is clear that an increase in k_2 results in shifting the resonant frequency, ω_r, beyond the natural frequency; nonetheless, its adverse effect (instability) is mitigated by a higher damping characteristic. Furthermore, the impact of raising the nonlinear viscous damping characteristic parameter, c_2, is demonstrated in **Figure 10**, which effectively dampens vibrations in the resonant region. At the same time, the vibration response in non-resonant regions (i.e., at low and high frequency ranges) remains unchanged.

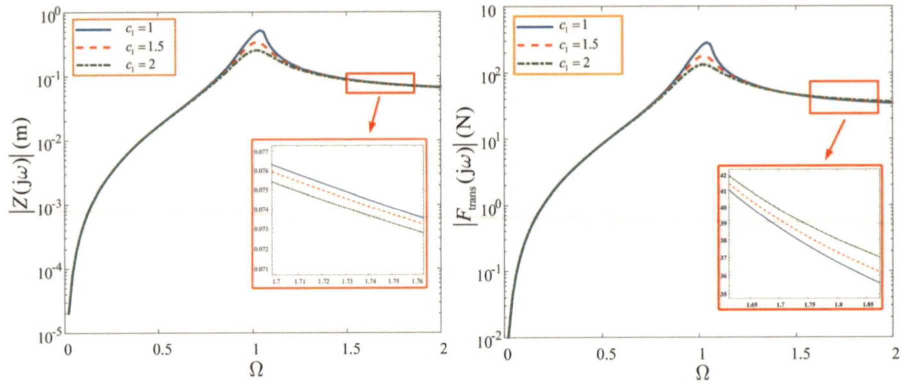

Figure 8.
Influence of linear damping on the system output responses, with $c_2 = 10\ N\ s\ m^{-3}$ and $k_2 = 100\ N\ m^{-3}$.

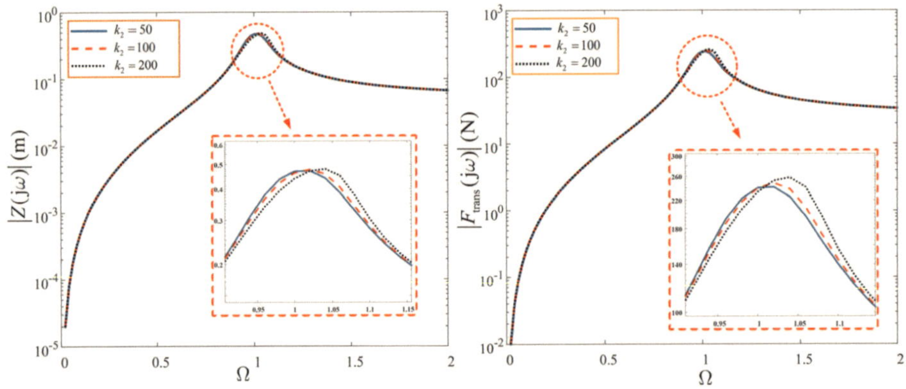

Figure 9.
Influence of nonlinear stiffness on the system output responses with $c_1 = 0.5\ N\ s\ m^{-1}$ and $c_2 = 10\ N\ s\ m^{-3}$.

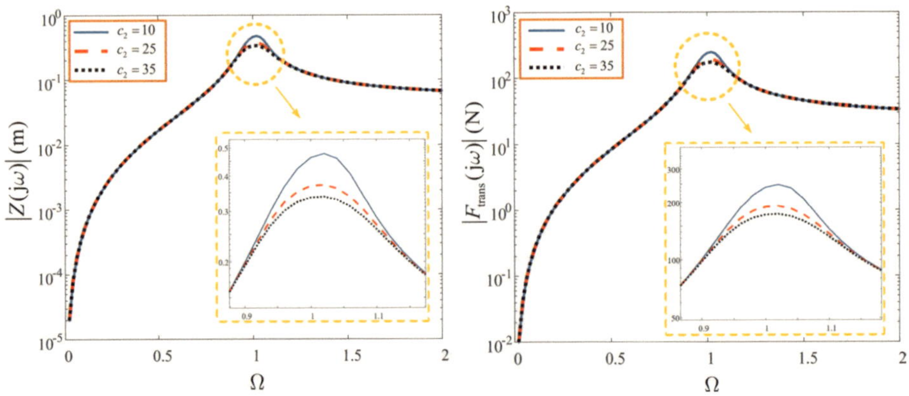

Figure 10.
Influence of nonlinear damping on the system output responses with $c_1 = 0.5\ N\ s\ m^{-1}$ and $k_2 = 100\ N\ m^{-3}$.

7. Energy dissipation capability of the suspension system

In **Figure 2**, the SDOF isolation system (spring and damper) loses energy as a result of its absorption and dissipative capabilities. The dissipated power at a time instance is given as $f_{\text{trans}}\dot{z}$, while the energy loss over a complete cycle, E_{d}, is given as

$$\begin{aligned} E_{\text{d}} &= \int_0^{\frac{2\pi}{\omega}} f_{\text{trans}}\dot{z}dt \\ &= \int_0^{\frac{2\pi}{\omega}} \left[\left(c_1\dot{z} + c_2 z^2\dot{z} + k_1 z + k_2 z^3 \right) \cdot \dot{z} \right] dt \end{aligned} \tag{19}$$

If the relative displacement $z = Z\sin(\omega t)$, then $\dot{z} = \omega Z\cos(\omega t)$. Eq. (19), therefore, becomes

$$E_{\text{d}} = \int_0^{\frac{2\pi}{\omega}} \left[\begin{pmatrix} c_1\omega Z\cos(\omega t) + c_2\omega Z^3\sin^2(\omega t)\cos(\omega t) \\ +k_1 Z\sin(\omega t) + k_2 Z^3\sin^3(\omega t) \end{pmatrix} \cdot \omega Z\cos(\omega t) \right] dt \tag{20}$$

which yields

$$\begin{aligned} E_{\text{d}} &= \pi c_1\omega Z^2 + \frac{1}{4}\pi c_2\omega Z^4 \\ &= \pi\omega Z^2\left(c_1 + \frac{1}{4}c_2 Z^2\right) \\ &= \pi\omega Z^2(c_1 + c_e) \\ &= \pi c_{eq}\omega Z^2 \end{aligned} \tag{21}$$

where $c_{eq} = c_1 + c_e$ and $c_e = \frac{1}{4}c_2 Z^2$.

It has been noted from Eq. (21) that the primary reason for energy dissipation in the suspension system is its damping properties. However, it is important to remember, as indicated in Eq. (16), that Z is influenced by both nonlinear damping c_2 and nonlinear stiffness, k_2. The overall frequency response function (OFRF) of the energy lost by the suspension system during one complete cycle can be obtained by replacing Eq. (15) in Eq. (21), resulting in

$$E_{\text{d}}(\omega, c_2, k_2) = \pi c_1\omega|Z(\mathrm{j}\omega)|^2 + \frac{1}{4}\pi c_2\omega|Z(\mathrm{j}\omega)|^4 \tag{22}$$

Eq. (22) indicates that E_{d} is associated with c_1, ω, Z, as well as the design parameters, c_2 and k_2. This shows that the OFRF model of the energy dissipated can be formulated in terms of the design parameters.

Similar to the OFRF model of the output response, $Z(\mathrm{j}\omega)$, of system (4), as deduced and demonstrated in Eq. (15), the OFRF models of the squared and quartic magnitudes of $Z(\mathrm{j}\omega)$, are also derived as

$$\begin{cases} |Z(\mathrm{j}\omega)|^2 = \sum\limits_{\bar{n}=0}^{\lfloor (N-1)/2 \rfloor} \sum\limits_{\bar{m}=0}^{\bar{n}} \mathrm{A}_{\bar{m},\bar{n}-\bar{m}}(\omega) c_2^{\bar{m}} k_2^{\bar{n}-\bar{m}}, \\ \text{and} \\ |Z(\mathrm{j}\omega)|^4 = \sum\limits_{\bar{n}=0}^{\lfloor (N-1)/2 \rfloor} \sum\limits_{\bar{m}=0}^{\bar{n}} \mathrm{B}_{\bar{m},\bar{n}-\bar{m}}(\omega) c_2^{\bar{m}} k_2^{\bar{n}-\bar{m}} \end{cases} \tag{23}$$

where $N = 7$, while $A_{\overline{m},\overline{n}-\overline{m}}$ and $B_{\overline{m},\overline{n}-\overline{m}}$ are frequency functions and denote the coefficients of the OFRF models, $|Z(j\omega)|^2$ and $|Z(j\omega)|^4$, respectively. Substituting Eq. (23) into Eq. (22) yields

$$E_d(\omega, c_2, k_2) = (\pi c_1 \omega) \cdot \sum_{\overline{n}=0}^{\lfloor (N-1)/2 \rfloor} \sum_{\overline{m}=0}^{\overline{n}} A_{\overline{m},\overline{n}-\overline{m}}(\omega) c_2^{\overline{m}} k_2^{\overline{n}-\overline{m}} + \left(\frac{1}{4}\pi\omega\right) \cdot \sum_{\overline{n}=0}^{\lfloor (N-1)/2 \rfloor} \sum_{\overline{m}=0}^{\overline{n}} B_{\overline{m},\overline{n}-\overline{m}}(\omega) c_2^{\overline{m}+1} k_2^{\overline{n}-\overline{m}} \tag{24}$$

Eq. (24) demonstrates that an estimate of the energy dissipated per cycle by the suspension system can be determined based on a specific set of design parameters. The relationship between energy dissipation and these design parameters, particularly at the resonant frequency, $\Omega = 1$ is illustrated in **Figure 11**. As shown in **Figure 12**, a substantial amount of energy is dissipated when a linear vibration isolator is employed ($c_2 = 0, k_2 = 0$). However, the energy dissipation decreases with the incorporation of nonlinear damping and stiffness characteristics. This trend can be attributed to the influence of nonlinear components on the relative displacement of the suspension system. With an increase in nonlinearities, the relative displacement diminishes, which in turn reduces both the force transmitted and the energy dissipated by the suspension system. It is also noteworthy that energy dissipation is highly sensitive to the nonlinear damping characteristics, while it is less responsive to the characteristics of nonlinear hardening stiffness. Surface plots illustrating the energy dissipated by the suspension system in relation to the design parameters beyond the resonant region, at $\Omega = 0.4$ and $\Omega = 2$, are presented in **Figures 13** and **14**, respectively.

Additionally, Eq. (24), representing the OFRF of energy dissipation in the suspension system, can be utilised to estimate energy dissipation for a specific set of optimal design parameters for c_2 and k_2. An estimation guide can be developed using an OFRF-

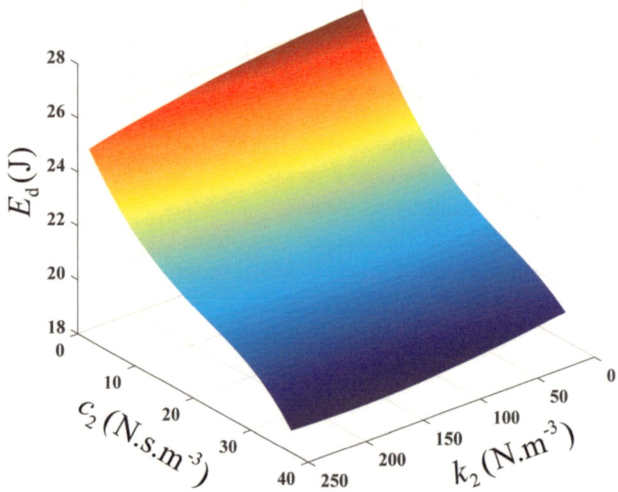

Figure 11.
Relationship between the energy dissipation and design parameters, at $\Omega = 1$.

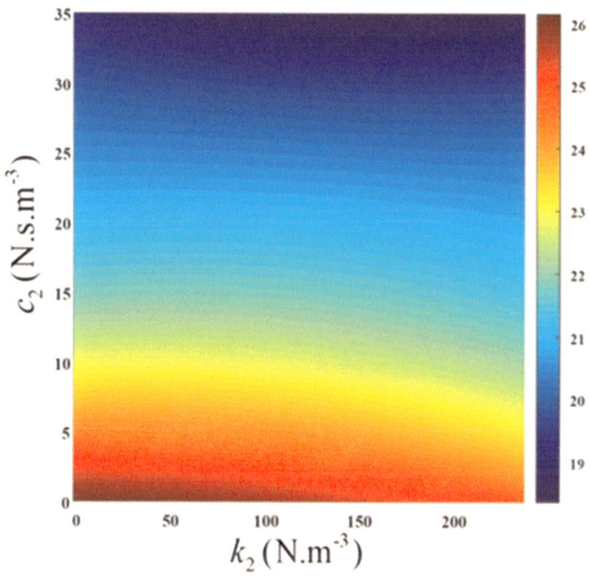

Figure 12.
Energy dissipation plot for sets of design parameters, at $\Omega = 1$.

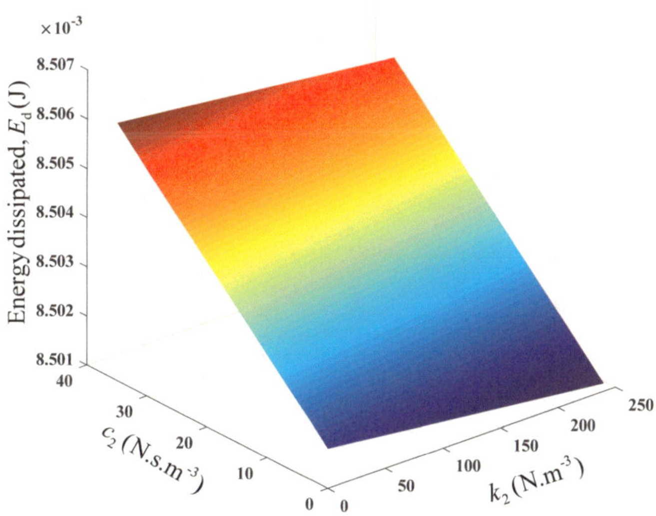

Figure 13.
Energy dissipation levels for different combinations of design parameters, at $\Omega = 0.4$.

generated contour map, which would assist in selecting design parameters aimed at achieving a specified energy dissipation level, as illustrated in **Figure 15**.

Choosing $k_2 = 200$ N m^{-3}, along the contour line of 20 J, gives a corresponding nonlinear damping value of $c_2 = 27.5$ N s m^{-3}. Substituting c_2 and k_2 into system (4) and numerically solving for the energy dissipation level, gives 21.08 J, which agrees with the design requirement having a percentage error of -5.1%.

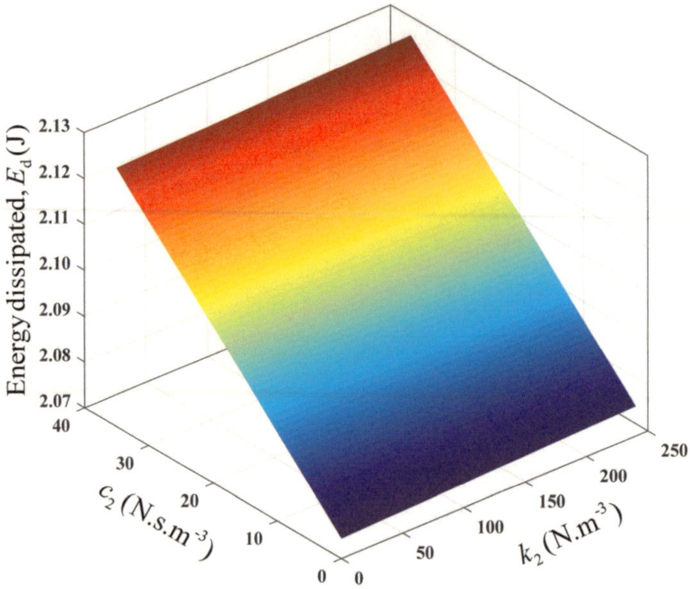

Figure 14.
Energy dissipation levels for different combinations of design parameters, at $\Omega = 2$.

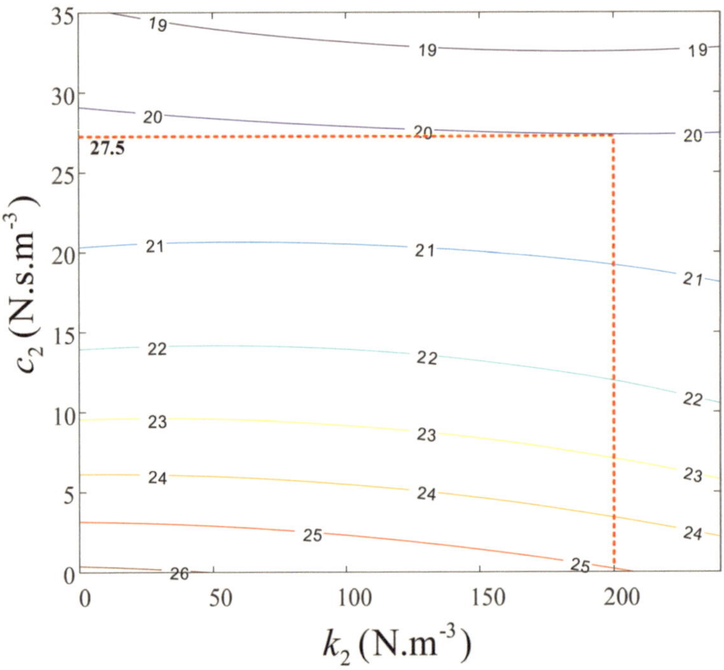

Figure 15.
Contours of energy dissipation levels, for sets of design parameters, at $\Omega = 1$.

The results from numerical studies to show the individual effects of the system parameters, c_1, c_2 and k_2, on the energy dissipation level of the suspension system, over the entire spectrum, are presented in **Figures 16–18**. This was done by varying either one of the parameters, c_1, c_2 or k_2, while keeping the other two fixed. In **Figure 16**, the negative effect of linear damping in the high-frequency range, observed in the output spectra of system (4), is also evident in the energy dissipation spectra. It is observed that while higher linear damping decreases energy dissipation in the resonant region, it considerably increases energy dissipation in the non-resonant

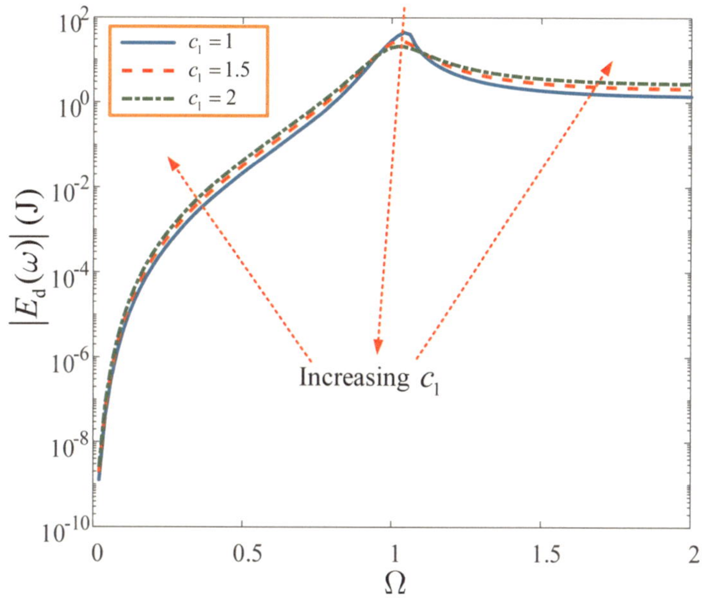

Figure 16.
Energy dissipation against frequency for different linear damping, with $c_2 = 10\ N\,s\,m^{-3}$ and $k_2 = 100\ N\,m^{-3}$.

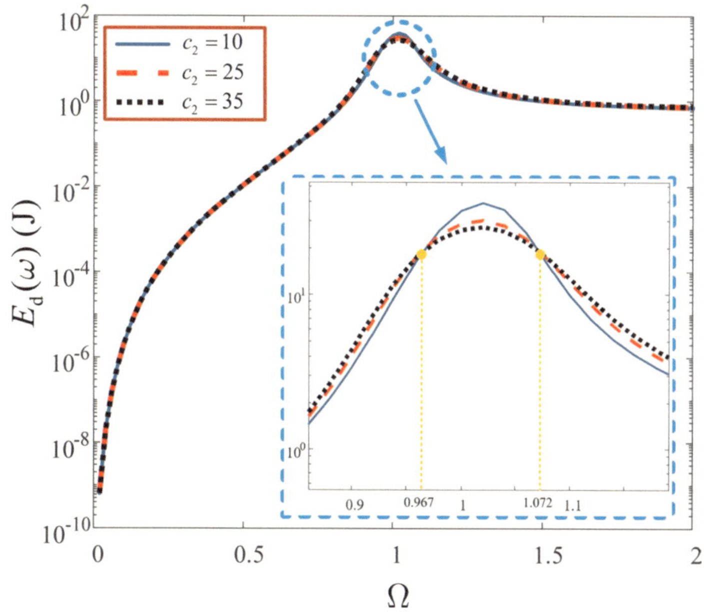

Figure 17.
Energy dissipation against frequency for different nonlinear damping, with $c_1 = 0.5\ N\,s\,m^{-1}$ and $k_2 = 100\ N\,m^{-3}$.

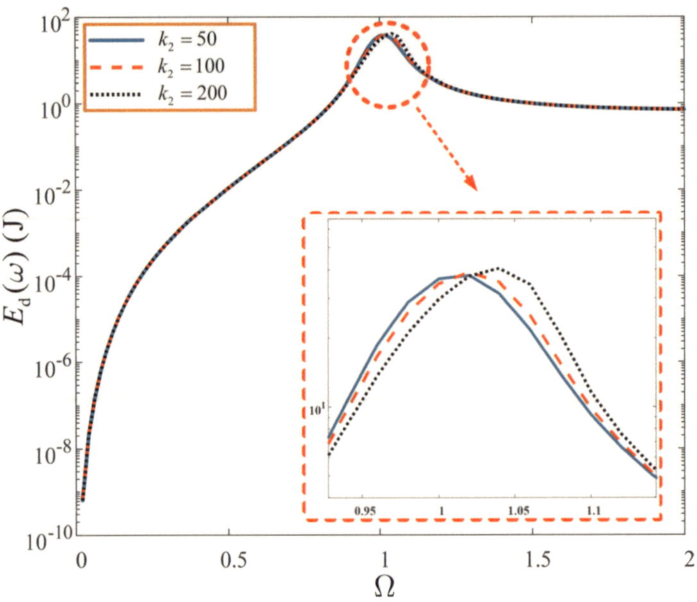

Figure 18.
Energy dissipation against frequency for different nonlinear stiffness, with $c_1 = 0.5\ N\,s\,m^{-1}$ *and* $c_2 = 10\ N\,s\,m^{-3}$.

regions. Similarly, an increase in nonlinear damping reduces energy dissipation in the resonant region. However, in contrast to linear damping, the corresponding rise in energy dissipation in the non-resonant regions is much less pronounced, as shown in **Figure 17**.

The influence of hardening stiffness, as shown in **Figure 18**, resembles the effect it has on the output spectra of system (4). In this case, as nonlinear stiffness is increased, the resonant peak of the energy dissipation is reduced at $\Omega = 1$. This is due to the resonant-shift effect of k_2, on the resonant energy, beyond the natural frequency.

By utilising the deduced OFRF models, **Figure 19** shows the disparities of both the energy dissipated and force transmitted by the suspension system as a function of the nonlinear stiffness parameter, k_2. One can see that the force transmitted by the suspension system is less sensitive to k_2, while the sensitivity of the energy dissipated to k_2, is negligible. This confirms the deduction made from **Figure 12**. This assertion primarily arises from the fact that the spring component retains potential energy when it is compressed or stretched, and loses negligible amounts of energy. It is also seen that for a fixed c_2, increasing k_2, causes an insignificant decrease in the energy dissipated per cycle. However, the vibration force transmitted is increased. In the same way, the findings represented in **Figure 20** show the changes in both energy dissipation and force transmission by the suspension system as they relate to the nonlinear damping parameter, c_2. It is clear in this instance that both performance metrics are highly sensitive to the parameter, c_2. Additionally, both metrics tend to decline progressively as the nonlinear damping parameter increases. This decline occurs because the relative displacement of the vehicle's suspension system decreases as the parameter, c_2 increases, leading to a reduction in both the force transmitted and the energy dissipated by the suspension system.

Figure 21 shows that the damping force F_d (N), peaks at the same frequency ($\Omega = 1$) where the energy dissipated,E_d (J), is maximum. This confirms that at the

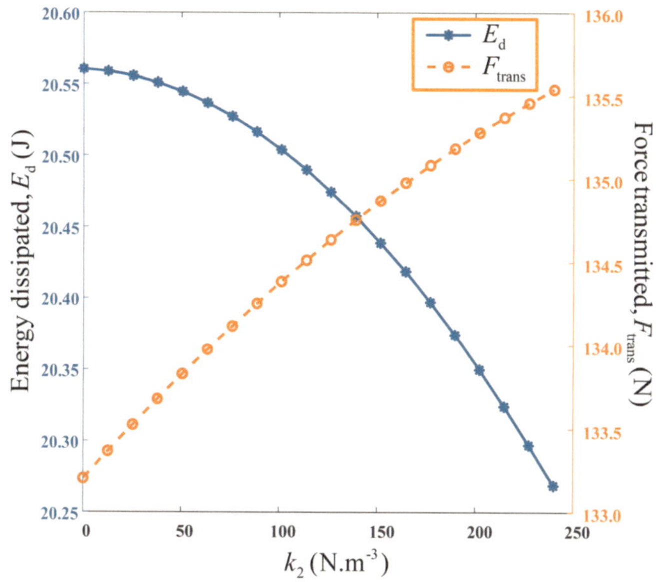

Figure 19.
Transmitted force and dissipated energy against nonlinear stiffness k_2 at $c_1 = 1.5\, N\, s\, m^{-1}$ and $c_2 = 23.9\, N\, s\, m^{-3}$.

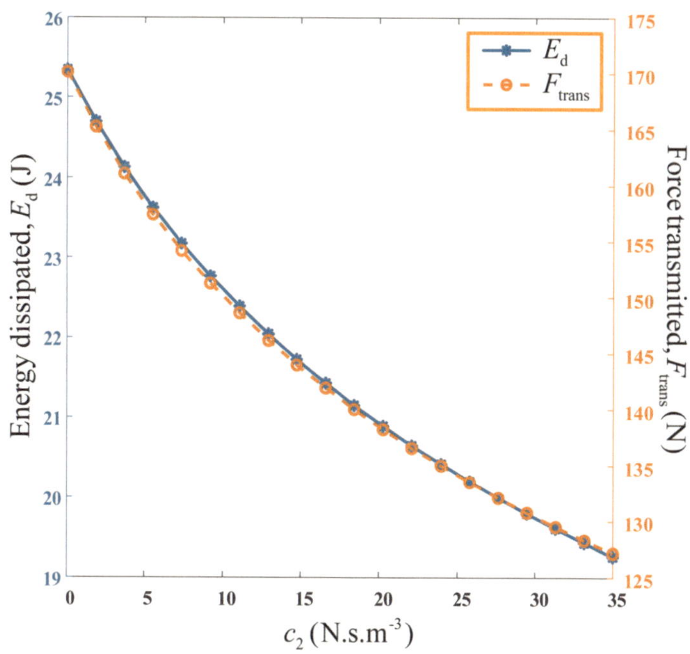

Figure 20.
Transmitted force and dissipated energy against nonlinear damping, c_2 at $c_1 = 1.5\, N\, s\, m^{-1}$ and $k_2 = 164.2\, N\, m^{-3}$.

maximum damping force, the maximum energy is dissipated by the suspension system.

This is expected, as a high damping force is necessary at resonance (where the displacement is at its maximum), which corresponds to a significant level of energy

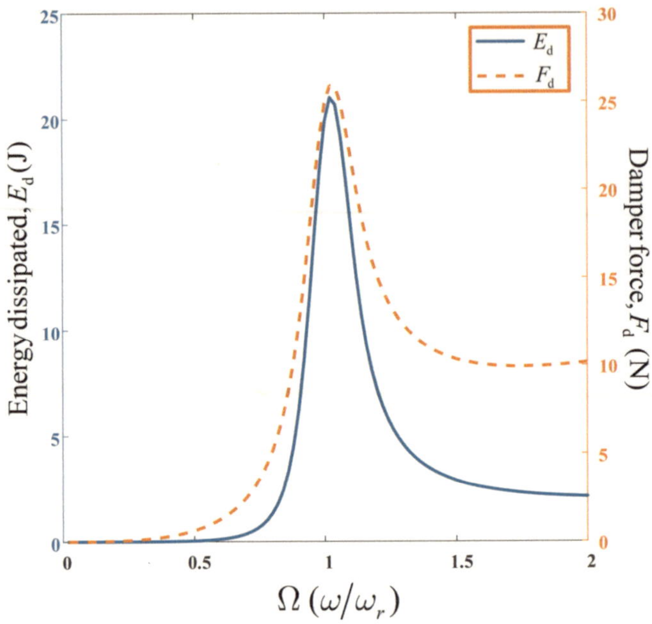

Figure 21.
Nonlinear damping and dissipated energy versus normalised frequency at $c_1 = 0.5\ N\,s\,m^{-1}$ *and* $c_2 = 10\ N\,s\,m^{-3}$.

dissipation by the suspension system. However, beyond the resonant frequency, the energy dissipation is considerably low in comparison to the level of damping force within other frequency range. This can be attributed to the minimal relative displacement experienced at high frequencies. In **Figure 22**, four hysteretic curves (A, B, C, and D) are presented, derived from different combinations of the designed nonlinear parameters of the system. For a vibration isolation system, the area under the force-displacement curve ($F_{trans} - Z$), indicates the energy dissipated per cycle by the isolation system at any frequency of interest, typically at the resonant frequency. It is noted that hysteretic curves C and D exhibit a smaller displacement span compared to curves A and B. This difference is clearly due to the effect of the nonlinear damping characteristics integrated into the suspension systems of C and D, whereas A and B only possess linear damping characteristics. Nonlinear dampers demonstrate superior vibration isolation performance compared to their linear counterparts (when $c_2 = 0$). The nonlinear damping, results in the decrease in the magnitude of the relative displacement, Z.

It is important to note that, in comparison to linear dampers, nonlinear dampers dissipate significantly more heat energy for an equal amount of excitation [29]. However, this comparison is not analysed in the current context. Additionally, while the hysteretic curve for a linear suspension system typically assumes an elliptical shape, this is not the case for a nonlinear system.

As observed in **Figure 23**, the effect of c_2, on curves C and D, with intercept $c_1 Z_{C,D}\omega_r = 18.75$N, on the vertical axis, is apparent, compared to the curves A and B (with intercept, $c_1 Z_{A,B}\omega_r = 24.96$N), without the effect of c_2. This is as a result of the influence of c_2, on the system output response, Z, which is a function of the nonlinear damping, c_2.

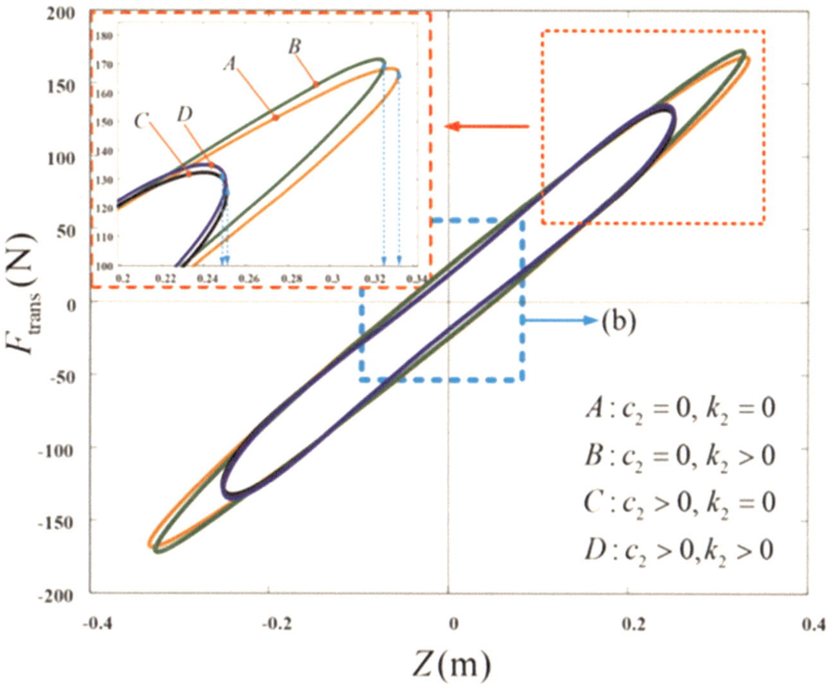

Figure 22.
*Comparative analysis of the Hysteretic curves of the suspension system with different combinations of nonlinear parameters, at resonance. (b) is shown in **Figure 23**.*

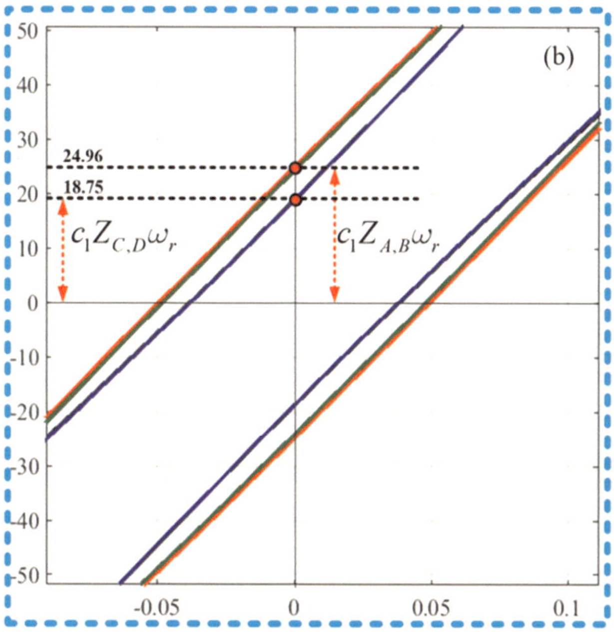

Figure 23.
*Enlarged origin of the hysteresis curves in **Figure 22**.*

8. Conclusions

In this study, we explored the vibration isolation performance of a nonlinear vehicle suspension system, modelled as a single-degree-of-freedom (SDOF) vibration isolation system. The Output Frequency Response Function (OFRF) method was employed as the primary analytical tool for system analysis and design, chosen for its ability to clearly express performance metrics in relation to key design parameters.

Both analytical and simulation results confirmed the effectiveness of this method in evaluating and designing vibration isolation systems. To emphasise the influence of nonlinear system parameters, we conducted a comparative study between the nonlinear model and its linear counterpart (where nonlinear components were set to zero). The findings showed that system nonlinearities significantly impact performance, particularly in the resonant region, where the peak resonance is notably reduced without compromising performance at higher frequencies. This highlights the superior vibration isolation capability of nonlinear suspension systems over linear ones.

Additionally, we analysed the energy dissipation per cycle of the designed suspension system using the OFRF technique. Through this approach, we established a polynomial relationship between the energy dissipation level and the system's nonlinear design parameters, enabling estimation for specific parameter sets. The results further reinforced the advantages of nonlinear suspension systems.

Moreover, numerical studies were conducted to examine the influence of the nonlinear parameters on the amount of energy dissipated. Lastly, the hysteretic curves were analysed to investigate how nonlinear system parameters influence the hysteresis loop shape. The findings revealed that energy dissipation is more sensitive to the system's nonlinear damping than to its stiffness. Overall, this study reiterates the vibration attenuation benefits of nonlinear suspension systems compared to the linear ones.

Author details

Uchenna Diala[1*] and Godwin Sani[2]

1 School of Engineering, University of Derby, Derby, United Kingdom

2 Deptartment of Computer Engineering, University of Calabar, Nigeria

*Address all correspondence to: u.diala@derby.ac.uk

References

[1] Cai LW. Fundamentals of Mechanical Vibrations. 1st edition. John Wiley & Sons, Wiley-ASME Press Series; April 2016. ISBN: 978-1-119-05022-3 Available from: https://www.wiley.com/en-us/Fundamentals+of+Mechanical+Vibrations-p-9781119050223

[2] Borowiec M, Sen AK, Litak G, Hunicz J, Koszałka G, Niewczas A. Vibrations of a vehicle excited by real road profiles. Forschung im Ingenieurwesen. 2010;**74**(2):99-109. DOI: 10.1007%2Fs10010-010-0119-y

[3] Soliman JI, Ismailzadeh E. Optimization of unidirectional viscous damped vibration isolation system. Journal of Sound and Vibration. 1974; **36**(4):527-539. DOI: 10.1016/S0022-460X(74)80119-7

[4] Liu M, Gu F, Zhang Y. Ride comfort optimisation of in-wheel-motor electric vehicles with in-wheel vibration absorbers. Energies. 2017;**10**(10):1647. DOI: 10.3390/en10101647

[5] Arana C, Evangelou SA, Dini D. Series active variable geometry suspension for road vehicles. IEEE/ASME Transactions on Mechatronics. 2014;**20**(1):361-372. DOI: 10.1109/TMECH.2014.2324013

[6] Alleyne A, Karl Hedrick J. Nonlinear adaptive control of active suspensions. IEEE Transactions on Control Systems Technology. 1995;**3**(1):94-101. DOI: 10.1109/87.370714

[7] Deringöl AH, Güneyisi EM. Influence of nonlinear fluid viscous dampers in controlling the seismic response of the base-isolated buildings. In: Structures. Vol. 34. Elsevier; 2021. pp. 1923-1941. Available from: https://www.sciencedirect.com/journal/structures. DOI: 10.1016/j.istruc.2021.08.106

[8] Kumar S, Medhavi A, Kumar R. Optimization of nonlinear passive suspension system to minimize road damage for heavy goods vehicle. International Journal of Acoustics and Vibrations. 2021;**26**(1):56-63. Available from: https://link.gale.com/apps/doc/A659376667/AONE?u=anon~b7686b90&sid=googleScholar&xid=fdd7cfc4

[9] Tiwari V, Sharma SC, Harsha SP. Ride comfort analysis of high-speed rail vehicle using laminated rubber isolator based secondary suspension. Vehicle System Dynamics. 2023;**61**(10):2689-2715. DOI: 10.1080/00423114.2022.2131584

[10] Lu Z, Wang Z, Zhou Y, Lu X. Nonlinear dissipative devices in structural vibration control: A review. Journal of Sound and Vibration. 2018; **423**:18-49. DOI: 10.1016/j.jsv.2018.02.052

[11] He S, Chen K, Xu E, Wang W, Jiang Z. Commercial vehicle ride comfort optimization based on intelligent algorithms and nonlinear damping. Shock and Vibration. 2019; **2019**(1):2973190. DOI: 10.1155/2019/2973190

[12] Wei C. Design and analysis of a novel vehicle-mounted active QZS vibration isolator. Iranian Journal of Science and Technology, Transactions of Mechanical Engineering. 2023;**47**(4): 2121-2131. DOI: 10.1007/s40997-023-00622-4

[13] Suman S, Balaji PS, Selvakumar K, Kumaraswamidhas LA. Nonlinear vibration control device for a vehicle suspension using negative stiffness mechanism. Journal of Vibration

Engineering & Technologies. 2021;**9**: 957-966. DOI: 10.1007/s42417-020-00275-6

[14] Shafiei B. Investigating the impact of the velocity of a vehicle with a nonlinear suspension system on the dynamic behavior of a Bernoulli–Euler bridge. SN Applied Sciences. 2021;**3**(3):293. DOI: 10.1007/s42452-021-04280-6

[15] Fernandes JCM, Gonçalves PJP, Silveira M. Interaction between asymmetrical damping and geometrical nonlinearity in vehicle suspension systems improves comfort. Nonlinear Dynamics. 2020;**99**(2):1561-1576. DOI: 10.1007/s11071-019-05374-y

[16] Borowiec M, Litak G, Friswell MI. Nonlinear response of an oscillator with a magneto-rheological damper subjected to external forcing. Applied Mechanics and Materials. 2006;**5**:277-284. DOI: 10.4028/www.scientific.net/AMM.5-6.277

[17] Jones JP. Automatic computation of polyharmonic balance equations for nonlinear differential systems. International Journal of Control. 2003; **76**(4):355-365. DOI: 10.1080/0020717031000079436

[18] Elliott AM, Bernstein MA, Ward HA, Lane J, Witte RJ. Nonlinear averaging reconstruction method for phase-cycle SSFP. Magnetic Resonance Imaging. 2007;**25**(3):359-364. DOI: 10.1016/j.mri.2006.09.013

[19] Stanzhitskii AN, Dobrodzii TV. Study of optimal control problems on the half-line by the averaging method. Differential Equations. 2011;**47**:264-277. DOI: 10.1134/S0012266111020121

[20] Xing XN, Zhang DM. Realization of nonlinear describing function method virtual experiment system based on LabVIEW. In: International Conference of China Communication (ICCC2010). Scientific Research Books; 2010. Available from: https://www.scirp.org/(S(351jmbnt-vnsjt1aadkozje))/proceeding/detailedinforofaproceeding?bookid=1285&booktypeid=3

[21] Attari M, Haeri M, Tavazoei MS. Analysis of a fractional order Van der Pol-like oscillator via describing function method. Nonlinear Dynamics. 2010;**61**: 265-274. DOI: 10.1007/s11071-009-9647-0

[22] Lang ZQ, Billings SA, Yue R, Li J. Output frequency response function of nonlinear Volterra systems. Automatica. 2007;**43**(5):805-816. DOI: 10.1016/j.automatica.2006.11.013

[23] Jing XJ, Lang ZQ, Billings SA. Output frequency response function-based analysis for nonlinear Volterra systems. Mechanical Systems and Signal Processing. 2008;**22**(1):102-120. DOI: 10.1016/j.ymssp.2007.06.010

[24] Jing XJ, Lang ZQ, Billings SA, Tomlinson GR. The parametric characteristic of frequency response functions for nonlinear systems. International Journal of Control. 2006; **79**(12):1552-1564. DOI: 10.1080/00207170600861019

[25] Jazar GN, Houim R, Narimani A, Golnaraghi MF. Frequency response and jump avoidance in a nonlinear passive engine mount. Journal of Vibration and Control. 2006;**12**(11):1205-1237. DOI: 10.1177/1077546306068059

[26] Peng ZK, Lang ZQ. The effects of nonlinearity on the output frequency response of a passive engine mount. Journal of Sound and Vibration. 2008; **318**(1-2):313-328. DOI: 10.1016/j.jsv.2008.04.016

[27] Milovanovic Z, Kovacic I, Brennan MJ. On the displacement transmissibility of a base excited viscously damped nonlinear vibration isolator. Journal of Vibration and Acoustics. 2009;**131**(5):1-054502. DOI: 10.1115/1.3147140

[28] Zhu Y, Lang ZQ. Design of nonlinear systems in the frequency domain: An output frequency response function-based approach. IEEE Transactions on Control Systems Technology. 2017; **26**(4):1358-1371. DOI: 10.1109/TCST.2017.2716379

[29] Denoël V, Degée H. Analysis of linear structures with nonlinear dampers. In: Eurodyn. ORBi, the institutional repository of the University of Liège; 2002. Available from: https://orbi.uliege.be/handle/2268/13026; https://orbi.uliege.be/

Chapter 3

Structural Stability of Materials

Amir Aliakbari and Peiman Amiri

Abstract

While experimental synthesis is essential for confirming the stability of materials, it's equally important to examine their structural stability from a theoretical standpoint. This theoretical analysis can reinforce practical synthesis efforts, whether we are dealing with bulk materials or two-dimensional substances. In this chapter, we'll look at different aspects of material stability, including dynamical, mechanical, and thermodynamical stability. We'll start by discussing phonon dispersion, which gives us insight into dynamical stability. For mechanical stability, we'll focus on elastic constants under ambient pressure conditions. Finally, in the context of thermodynamical stability, we'll explain cohesive energy, which represents the energy needed to break atomic bonds in a solid and turn them into isolated atoms.

Keywords: stability of materials, theoretical analysis, dynamical stability, mechanical stability, thermodynamical stability

1. Introduction

Understanding how materials hold up under different conditions is a key concept in materials science and engineering. The ability of a material to maintain its structure and integrity when faced with external factors is crucial for determining its suitability for various applications [1]. Structural stability goes beyond being physically strong; it involves a complex interplay of dynamic, mechanical, and thermodynamic factors. Recently, the significance of grasping structural stability has grown tremendously, especially with the rise of innovative materials like 2D structures, metamaterials, and complex composites. These newer materials often showcase distinct stability characteristics that differ greatly from traditional bulk materials. For instance, graphene, which is incredibly thin at just one atom thick, demonstrates outstanding structural stability due to its unique bonds and lattice structure [2]. Analyzing structural stability involves examining different levels of detail. At the atomic level, we need to consider the type and strength of chemical bonds, which directly impact a material's ability to hold together. Moving to a smaller scale, crystal structure and defects become critical. At the macroscopic level, properties like elastic constants and how materials expand with temperature changes come into play [3]. A significant challenge in modern materials science is predicting a material's stability before it is made. This predictive capability has become increasingly important as material design shifts toward computational methods and AI-driven discovery. Theoretical frameworks for analyzing stability have evolved from simple rules to advanced quantum calculations and simulations using molecular dynamics [4]. Modern theoretical approaches to structural stability typically involve three main aspects:

1. Dynamical stability: Analysis of phonon spectra and vibrational modes

2. Mechanical stability: Evaluation of elastic constants and their relationships

3. Thermodynamical stability: Assessment of energy landscapes and phase transitions

Each of these factors offers distinct perspectives on a material's overall stability characteristics, aiding in forecasting how it will perform in different situations. For example, conducting dynamical stability assessments using phonon calculations can highlight potential instabilities that may not be evident solely through examining the material's structure [5, 6]. The implications of structural stability reach far beyond academic curiosity. In engineering applications, a deep comprehension of stability is vital for:

- Predicting material failure modes

- Designing more durable structures

- Optimizing processing conditions

- Developing new materials with enhanced properties

- Ensuring long-term reliability in service conditions

2. Theoretical foundations

The theoretical understanding of structural stability begins with fundamental principles of solid-state physics and materials science. This section establishes the mathematical and physical frameworks necessary for analyzing material stability across different scales and conditions.

2.1 Energy considerations in crystal structures

The complete energy of a crystalline system can be represented as a function of both atomic positions and lattice parameters. In order for a crystal structure to be considered stable, it should correspond to a local minimum on the potential energy surface. This condition can be mathematically articulated as [7]:

$$\mathbf{E} = \mathbf{E}(\mathbf{R_1}.\mathbf{R_2}.\cdots.\mathbf{R_n}.\mathbf{V}) \tag{1}$$

where $\mathbf{E}$ is the total energy, $\mathbf{R_i}$ represents atomic positions, and $\mathbf{V}$ is the volume. For stability, the following conditions must be satisfied:

$$\partial\mathbf{E}/\partial\mathbf{R_i} = \mathbf{0} \text{ (equilibrium condition)} \tag{2}$$

$$\partial^2 \mathbf{E}/\partial \mathbf{R}_i^2 > \mathbf{0} \text{ (stability condition)} \tag{3}$$

The Born-Huang criteria for stability provide a set of conditions that elastic constants in a crystal structure must meet to ensure the stability of the material. These criteria outline mathematical relationships that the elastic constants need to satisfy, and these relationships differ depending on the specific crystal system under consideration [8].

2.2 Quantum mechanical foundations

Density functional theory (DFT) calculations are a prevalent method in modern stability analysis, offering a practical approach to solving the complex many-body Schrödinger equation. This technique provides a computationally feasible means of expressing the total energy of a system, aiding in the assessment of material stability and properties. The total energy in DFT is expressed as [9]:

$$\mathbf{E}[\boldsymbol{\rho}] = \mathbf{T}[\boldsymbol{\rho}] + \mathbf{V_{ext}}[\boldsymbol{\rho}] + \mathbf{V_H}[\boldsymbol{\rho}] + \mathbf{E_{xc}}[\boldsymbol{\rho}] \tag{4}$$

where $\boldsymbol{T}[\rho]$ is the kinetic energy functional, $\boldsymbol{V_{ext}}[\rho]$ represents external potential energy, $\boldsymbol{V_H}[\rho]$ is the Hartree potential, and $\boldsymbol{E_{xc}}[\rho]$ is the exchange-correlation functional.

2.3 Statistical mechanical approach

Temperature effects on material stability are studied using statistical mechanics, with the Gibbs free energy serving as a pivotal quantity for understanding the stability of systems at different temperatures. This parameter accounts for the balance between the enthalpy and entropy of a system, providing valuable insights into the thermodynamic stability of materials. The Gibbs free energy, crucial for stability analysis, is given by Refs. [10–12]:

$$\mathbf{G} = \mathbf{H} - \mathbf{TS} = \mathbf{U} + \mathbf{PV} - \mathbf{TS} \tag{5}$$

Where **H** is enthalpy, **T** is temperature, **S** is entropy, **U** is internal energy, **P** is pressure, and **V** is volume.

2.4 Continuum mechanics framework

In continuum mechanics, stability analysis investigates material behavior on a large scale. The generalized Hooke's law describes how stress ($\boldsymbol{\sigma_{ij}}$) and strain ($\boldsymbol{\varepsilon_{kl}}$) are related through the elastic stiffness tensor ($\mathbf{C_{ijkl}}$), offering a framework for understanding the mechanical response of materials under various loading conditions [13]:

$$\boldsymbol{\sigma_{ij}} = \mathbf{C_{ijkl}}\boldsymbol{\varepsilon_{kl}} \tag{6}$$

For stability, the strain energy density must be positive definite:

$$\mathbf{U} = \frac{\mathbf{1}}{\mathbf{2}}\mathbf{C_{ijkl}}\boldsymbol{\varepsilon_{ij}}\boldsymbol{\varepsilon_{kl}} \tag{7}$$

2.5 Computational methods

Theoretical studies on structural stability heavily rely on cutting-edge computational tools, enabling a deeper understanding of material behaviors. These modern computational approaches, such as DFT, DFPT, MD, and Monte Carlo simulations, offer valuable insights into the stability and properties of materials under various conditions:

1. Density Functional Theory (DFT): Used for electronic structure and cohesive energy calculations.
2. Density Functional Perturbation Theory (DFPT): An extension of DFT, utilized for phonon dispersion and vibrational mode analyses.
3. Molecular Dynamics (MD): Simulates atomic-scale dynamics to study temperature effects on structural integrity.
4. Monte Carlo Simulations: Effective for evaluating phase stability and critical temperature of phase transitions.

Key software packages include:

- VASP (Vienna Ab Initio Simulation Package): Widely used for DFT and MD simulations.
- Quantum ESPRESSO: Known for DFPT calculations.
- LAMMPS (Large-scale Atomic/Molecular Massively Parallel Simulator): Focuses on MD simulations.

For example, Quantum ESPRESSO studies have unveiled phonon dispersion trends in perovskites and layered materials, aiding in precise dynamical stability evaluations [14, 15]. Likewise, LAMMPS simulations have forecasted the temperature-dependent mechanical stability of metallic alloys [16]. The selection of the computational method is guided by the scale of the analysis and the level of accuracy needed for the investigation [17].

3. Dynamical stability

Exploring dynamical stability is crucial in understanding how materials withstand small disruptions, particularly by studying vibrational properties like phonons that reveal atomic responses [6, 18]. This section intricately examines the computational and theoretical systems that characterize phonon dispersion, lattice dynamics, and temperature impacts, emphasizing their importance in evaluating material stability. By delving into these realms, researchers can gain profound insights into how materials react at the atomic scale in diverse scenarios, enriching our comprehension of structural integrity and stability forecasts.

3.1 Phonon dispersion

Understanding phonon dispersion is crucial for assessing a material's propensity for dynamic stability and predicting potential structural transformations. Phonon dispersion elucidates the connection between phonon frequencies and wave vectors in a crystalline lattice [5]. A material's dynamical stability is dictated by its phonon spectrum; the detection of imaginary phonon frequencies at certain wave vectors indicates dynamic instability, often signaling a predisposition toward structural rearrangements or phase transitions [19].

Key insights from phonon dispersion analysis:

1. Structural stability checks: Phonon dispersion curves offer valuable insights into structural viability. Notably, cubic perovskites like $BaTiO_3$ showcase imaginary phonon modes at reduced temperatures, indicating a predisposition toward transitioning to phases with lower symmetry [20].

2. Phase transition prediction: For example, the YC in both the NaCl (B1) phase and the CsCl (B2) phase was extensively studied using density functional theory (DFT). In the **Figure 1** below (left), it can be seen that for the compound YC in the B1 phase at ambient pressure, there is no imaginary phonon frequency in the entire Brillouin zone, so this compound is dynamically stable in this phase, and it is possible to synthesize this compound. However, the figure on the right side confirms the instability of the B2 phase at ambient pressure. Therefore, the synthesis of the compound YC in the B2 phase is impossible [6].

3. Defect effects: Phonon dispersion analyses extend to evaluating the impact of lattice defects, which can either stabilize or destabilize the structure. In graphene, the introduction of vacancies alters phonon modes, consequently impacting thermal conductivity and mechanical robustness [21].

Sophisticated computational tools such as Quantum ESPRESSO, PHONOPY, and VASP have revolutionized phonon calculations, aiding in the design of thermally robust 2D structures like MXenes and bulk materials [5, 14]. These tools are instrumental in predicting and optimizing the stability and properties of materials, offering valuable insights for material engineers and researchers.

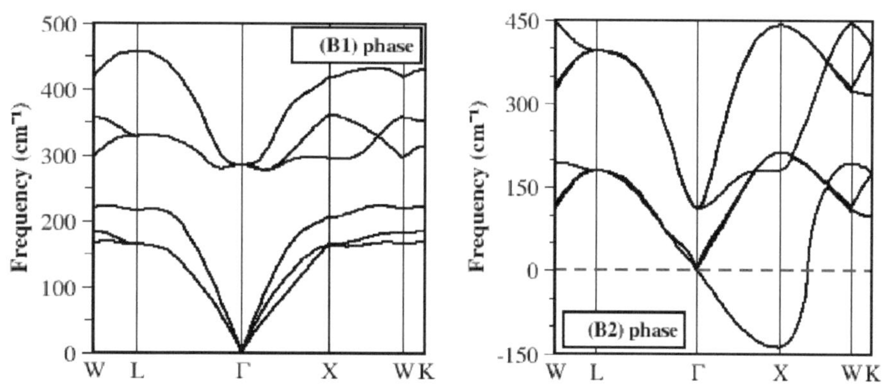

Figure 1.
Phonon dispersion spectrum of YC structure within the B1 phase (left side), and for the B2 phase (right side) at ambient pressure [6].

3.2 Lattice dynamics

Lattice dynamics concentrates on atomic vibrations (phonons) and their influence on a material's structural and thermal characteristics.

1. Harmonic approximation: This method simplifies vibrational analysis by presuming atomic displacements are small and controlled by quadratic potential energy terms. Although effective at low temperatures, it overlooks temperature-related effects such as thermal expansion. Nanostructured materials like zeolites and metal-organic compounds reveal remarkable consistency between harmonic model forecasts and experimental vibrational spectra under cryogenic conditions [22].

2. Anharmonic effects: Anharmonicity becomes notable at higher temperatures, impacting characteristics such as thermal expansion and phonon durations. The softening of optical phonon modes in lead halide perovskites is attributed to anharmonic lattice interactions, which drive phase transitions. For instance, the diminution of optical phonon modes in lead halide perovskites is linked to anharmonic lattice interactions, instigating phase transitions [23]. Anharmonic interactions in silicon nanowires significantly lower thermal conductivity compared to bulk silicon, making them suitable for thermoelectric applications [24, 25].

3. Coupling between phonons and electrons: Lattice vibrations impact electronic characteristics as well. In superconductors such as MgB_2, robust electron-phonon coupling bolsters superconducting phenomena, a phenomenon scrutinized through sophisticated lattice dynamics frameworks [26].

3.3 Temperature effects on dynamical stability

Temperature significantly impacts the vibrational stability of materials, influencing their structural integrity and behavior. Temperature variations can induce changes in phonon frequencies and lattice dynamics, leading to alterations in material properties.

1. Quasi-Harmonic Approximation (QHA): This computational approach accounts for thermal expansion and pressure influences, allowing for accurate forecasts of how a material's stability changes with temperature variations [6, 14, 15]. This computational tool is instrumental in understanding how materials behave under different thermal and pressure conditions, aiding in the prediction of their stability over a range of temperatures. Studies on boron arsenide (BAs) have demonstrated its high thermal conductivity and stability at extreme temperatures, making it ideal for heat dissipation in electronics [27].

2. Thermal Phase Transitions: In materials such as VO_2, shifts from an insulating monoclinic phase to a metallic rutile phase are triggered by temperature-induced lattice distortions. These phase transitions play a crucial role in applications like thermochromics and electronics [28, 29].

3. Debye Temperature and Stability (Θ_D): The Debye temperature is fundamental in understanding how materials transfer heat and respond to temperature

variations, providing insights into their thermal behavior and stability. This parameter plays a significant role in characterizing a material's ability to withstand thermal stresses and efficiently conduct heat. The following equation shows the Debye temperature, which describes the maximum phonon frequency (ω_D) of scattering curvs $\Theta_D = \frac{\hbar\omega_D}{k_B}$ is a key parameter linking lattice vibrations to thermal properties [6, 30]. Materials with high Debye temperatures, like diamond, signify robust structural integrity and excellent thermal conductivity, especially at elevated temperatures [31].

4. Mechanical stability

Mechanical stability is essential for ensuring the structural integrity of a material, enabling it to retain its shape and structure when subjected to external forces. Therefore, the evaluation of elastic parameters stands out as a crucial criterion in assessing the mechanical stability of materials [32, 33]. This section delves into both the theoretical underpinnings and practical considerations involved in the analysis of mechanical stability, shedding light on the intricate aspects of this critical assessment process.

4.1 Elastic constants and compliance

Elastic constants play a pivotal role in defining a material's mechanical response to external forces, governing its deformation behavior. Understanding these elastic constants is crucial for predicting a material's response to mechanical stress and determining its overall mechanical stability. Hooke's law, expressed in tensor notation, establishes the linear relationship between stress and strain in anisotropic materials, encapsulating their mechanical properties comprehensively [13].

$$\sigma_{ij} = C_{ijkl}\varepsilon_{kl} \tag{8}$$

By employing matrix notation, as detailed in Ref. [13], the four-suffix stiffness tensor (C_{ijkl}) can be condensed into a two-suffix stiffness tensor. This transformation involves merging the initial two suffixes into a unified suffix spanning from 1 to 6, a change also applied to the final two suffixes [13]. This method leverages the symmetry inherent in the second-rank stress and strain tensors, facilitating their representation as six-dimensional vectors within a coordinate system characterized by orthogonal axes. Voigt notation condenses σ_{ij} into a six-dimensional vector for easier representation and manipulation in mechanical analysis.

$$\sigma_{ij} = \begin{bmatrix} \sigma_{xx} & \sigma_{xy} & \sigma_{xz} \\ \sigma_{yx} & \sigma_{yy} & \sigma_{yz} \\ \sigma_{zx} & \sigma_{zy} & \sigma_{zz} \end{bmatrix} \tag{9}$$

$$\left(\sigma_{xx}.\sigma_{yy}.\sigma_{zz}.\sigma_{yz}.\sigma_{xz}.\sigma_{xy}\right) \equiv \left(\sigma_1.\sigma_2.\sigma_3.\sigma_4.\sigma_5.\sigma_6\right) \tag{10}$$

Similarly, ε_{ij} can be simplified to: $\left(\varepsilon_{xx}.\varepsilon_{yy}.\varepsilon_{zz}.\varepsilon_{yz}.\varepsilon_{xz}.\varepsilon_{xy}\right) \equiv \left(\varepsilon_1.\varepsilon_2.\varepsilon_3.\varepsilon_4.\varepsilon_5.\varepsilon_6\right)$.

Therefore, in Voigt notation, the relationship between stress and strain can be expressed concisely:

$$\begin{bmatrix} \sigma_1 \\ \sigma_2 \\ \sigma_3 \\ \sigma_4 \\ \sigma_5 \\ \sigma_6 \end{bmatrix} = \begin{bmatrix} C_{11} & C_{12} & C_{13} & C_{14} & C_{15} & C_{16} \\ C_{21} & C_{22} & C_{23} & C_{24} & C_{25} & C_{26} \\ C_{31} & C_{32} & C_{33} & C_{34} & C_{35} & C_{36} \\ C_{41} & C_{42} & C_{43} & C_{44} & C_{45} & C_{46} \\ C_{51} & C_{52} & C_{53} & C_{54} & C_{55} & C_{56} \\ C_{61} & C_{62} & C_{63} & C_{64} & C_{65} & C_{66} \end{bmatrix} \begin{bmatrix} \varepsilon_1 \\ \varepsilon_2 \\ \varepsilon_3 \\ \varepsilon_4 \\ \varepsilon_5 \\ \varepsilon_6 \end{bmatrix} \tag{11}$$

4.2 Bulk mechanical stability criteria

Understanding the $\mathbf{C_{ij}}$ matrix is crucial for assessing the mechanical stability of an unstrained crystal through the well-known Born-Huang mechanical stability criteria [8, 34, 35]. A crystal is considered mechanically stable only if the $\mathbf{C_{ij}}$ matrix is positive definite. While this criterion is necessary, it may not always be sufficient for various crystal systems. Reference [8] outlines the necessary and sufficient conditions for evaluating mechanical stability across all crystal classes. Additionally, the Space Group (SPG) identifiers corresponding to each crystal class are enumerated [8, 13, 34, 36].

4.2.1 Cubic structures: SPG#195–230

For cubic crystal structures, we find that only three independent $\mathbf{C_{ij}}$ coefficients, which are C_{11}, C_{12}, and C_{44} govern their mechanical stability. These structures adhere to specific mechanical stability criteria:

$$C_{11} - C_{12} > 0. \quad C_{44} > 0. \quad C_{11} + 2C_{12} > 0 \tag{12}$$

4.2.2 Hexagonal structures: SPG#168–194

Hexagonal crystal structures exhibit five independent $\mathbf{C_{ij}}$ coefficients C_{11}, C_{12}, C_{13}, C_{33}, and C_{44} that dictate their mechanical behavior. To ensure stability in such structures, C_{11}, C_{12}, C_{13}, and C_{33} must be all positive, along with the conditions:

$$C_{11} > 0. \; C_{11} - C_{12} > 0. \quad C_{44} > 0. \text{ and } \; (C_{11} + C_{12})C_{33} - 2(c_{13})^2 > 0 \tag{13}$$

4.2.3 Rhombohedral structures: SPG#143–167

Rhombohedral structures characterized by seven independent $\mathbf{C_{ij}}$ coefficients C_{11}, C_{12}, C_{13}, C_{14}, C_{15}, C_{33}, and C_{44} pose unique challenges for mechanical stability assessment. The specific mechanical stability conditions for these structures are:

$$C_{11} - C_{12} > 0. \; C_{44} > 0. C_{13}{}^2 < \frac{1}{2}(C_{11} + C_{12})C_{33} \; . \text{ and } \; C_{14}{}^2 + C_{15}{}^2 < \frac{1}{2}(C_{11} - C_{12})C_{44} \equiv C_{44}C_{66} \tag{14}$$

4.2.4 Tetragonal structures: SPG#75–142

Tetragonal crystal structures present two distinct classes: Class I (Laue class 4/mmm) with six independent coefficients and class II (Laue class 4/m) with seven.

Class I has six independent C_{ij} coefficients (C_{11}, C_{12}, C_{13}, C_{33}, C_{44}, and C_{66}), whereas class II has seven independent C_{ij} coefficients with $C_{16} \neq 0$. The mechanical stability requirements for each class of tetragonal structures are as below:

$$C_{11} - C_{12} > 0.\ C_{44} > 0.(C_{11} + C_{12})C_{33} > 2(C_{13})^2\ .\ \text{and}\ \ 2C_{16}^2 < (C_{11} - C_{12})C_{66} \tag{15}$$

4.2.5 Orthorhombic structures: SPG#16–74

Orthorhombic crystal structures feature nine independent C_{ij} coefficients that influence their mechanical stability. The exact mechanical stability criteria for the orthorhombic class can be expressed as follows:

$$C_{11} > 0.\ C_{44} > 0.C_{55} > 0.\ C_{66} > 0.C_{11}C_{22} > C_{12}^2 \tag{16}$$

$$C_{11}C_{22}C_{33} + 2C_{12}C_{13}C_{23} - C_{11}C_{23}^2 - C_{22}C_{13}^2 - C_{33}C_{12}^2 > 0 \tag{17}$$

4.2.6 Monoclinic structures: SPG#3–15

Monoclinic crystal structures possess 13 independent C_{ij} coefficients crucial for evaluating their mechanical stability. The general mechanical stability conditions for monoclinic structures include:

$$C_{11} + C_{22} + C_{33} + 2(C_{12} + C_{13} + C_{23}) > 0.C_{33}C_{55} - C_{35}^2 > 0.C_{44}C_{66} - C_{46}^2 > 0 \tag{18}$$

$$C_{22} + C_{33} - 2C_{23} > 0.C_{22}(.C_{33}C_{55} - C_{35}^2) + 2C_{23}C_{25}C_{35} - C_{55}C_{23}^2 - C_{33}C_{25}^2 > 0 \tag{19}$$

$$\begin{aligned}&2[C_{15}C_{25}(C_{33}C_{12} - C_{13}C_{23}) + C_{15}C_{35}(C_{22}C_{13} - C_{12}C_{23}) + C_{25}C_{35}(C_{11}C_{23} - C_{12}C_{13})] - \\ &[C_{15}^2(C_{33}C_{22} - C_{23}^2) + C_{25}^2(C_{33}C_{11} - C_{13}^2) + C_{35}^2(C_{11}C_{22} - C_{12}^2) + \\ &C_{55}(C_{33}C_{22}C_{11} - C_{11}C_{23}^2 - C_{22}C_{13}^2 - C_{33}C_{12}^2 + 2C_{12}C_{13}C_{23}) > 0\end{aligned} \tag{20}$$

4.2.7 Triclinic structures: SPG#1–2

Triclinic crystal structures, characterized by a comprehensive set of 21 nonzero independent C_{ij} coefficients, require intricate evaluations for mechanical stability. One can check the positive definiteness in the C_{ij} matrix serves as a fundamental check for ensuring the stability of triclinic structures, as expounded in scholarly works [8].

$$\begin{aligned}&C_{11} > 0.C_{12} > 0.C_{13} > 0.C_{14} > 0.\ C_{15} > 0.C_{16} > 0.C_{22} > 0.C_{23} > 0.C_{24} > 0.\ C_{25} > 0.C_{26} \\ &> 0.C_{33} > 0.C_{34} > 0.C_{35} > 0.C_{36} > 0.C_{44} > 0.C_{45} > 0.C_{46} > 0.C_{55} > 0.C_{56} > 0.C_{66} > 0.\end{aligned} \tag{21}$$

4.3 2D mechanical stability criteria

Using the C_{ij} matrix, the mechanical stability of a two-dimensional (2D) stress-free lattice can be assessed according to the criteria outlined in Ref. [37]. Like the three-dimensional (3D) case, the mechanical stability in two dimensions cannot be determined solely by verifying the positive definiteness of the C_{ij} matrix.

- Square lattice conditions:

$$\mathbf{C_{11} - C_{12} > 0. \quad C_{66} > 0. \quad C_{11} + C_{12} > 0} \tag{22}$$

- Rectangular class lattice conditions:

$$\frac{1}{2}\left(C_{11} + C_{12} + \sqrt{4C_{12}^{2} - (C_{11} - C_{22})^{2}}\right) > 0. \tag{23}$$

$$\frac{1}{2}\left(C_{11} + C_{12} - \sqrt{4C_{12}^{2} - (C_{11} - C_{22})^{2}}\right) > 0. \quad C_{66} > 0 \tag{24}$$

- Hexagonal lattice conditions:

$$\mathbf{C_{11} - C_{12} > 0. \quad C_{11} + C_{12} > 0} \tag{25}$$

5. Thermodynamical stability

Thermodynamic stability refers to a material's energy configuration to possible transformations, including phase transitions, decomposition, or defect formation. It offers valuable insights into the probability of a material retaining its structural integrity when exposed to different environmental conditions. This section highlights cohesive energy, formation energy, and phase stability as essential parameters for assessing thermodynamic stability.

5.1 Cohesive energy

Cohesive energy ($\mathbf{E_{coh}}$) is defined as the energy needed to break apart a solid into its individual free atoms, each in its ground state. This quantity serves as an indicator of the strength of atomic interactions within the crystal lattice. The cohesive energy per atom can be expressed mathematically as follows [38–40]:

$$\mathbf{E_{coh}} = -\frac{\mathbf{E_{tot}} - \sum \mathbf{E_{atomic}}}{\sum \mathbf{N}} \tag{26}$$

In the equation above, $\mathbf{E_{tot}}$ represents the total energy of the structure, $\mathbf{E_{atomic}}$ denotes the energy of a free atom, and N is the number of atoms within a unit cell. A high cohesive energy indicates strong atomic bonds and greater material stability. The calculated cohesive energies for various materials, including Sc_2C MXene, graphene, germanene, and silicene, are 6.48, 7.85, 3.26, and 3.98 eV/atom, respectively [41]. Among these, graphene stands out due to its exceptional hardness and high thermal stability. Furthermore, the differences in cohesive energy help explain why graphite is thermodynamically favored over diamond under standard conditions [42].

5.2 Formation energy

Formation energy ($\mathbf{E_f}$) measures the energy change associated with the formation of a compound from its constituent elements in their standard states [43]. It serves as a crucial parameter for assessing the stability of a compound and its tendency to undergo defect formation [44]. The formation energy is determined by the difference between the total energy of the compound and the energy required for its equilibrium decomposition into the constituent elements in their ground-state structural phases. For a given compound:

$$\mathbf{E_f} = \mathbf{E_{tot}} - \sum \mathbf{E_{decomposition}} \tag{27}$$

A negative formation energy indicates that the synthesis of a compound is thermodynamically favorable. For instance, the low formation energy of TiO_2 (-9.7 eV) contributes to its stability, making it a suitable material for applications such as photocatalysis and coatings [45]. Additionally, the formation energies of defects, including vacancies, interstitials, and substitutions, play a critical role in determining defect concentrations and their influence on the material's properties [46].

5.3 Phase stability

Phase stability refers to a material's capacity to maintain a specific phase under varying conditions of temperature, pressure, or composition [47]. The stability of a phase is typically assessed using Gibbs free energy (**G**), which is minimized for the thermodynamically stable phase [10–12]. Gibbs free energy is expressed as:

$$\mathbf{G} = \mathbf{H} - \mathbf{TS} \tag{28}$$

where **H** is enthalpy, **T** is temperature, and **S** is entropy.

5.3.1 Key techniques

1. Phase diagrams: Computational phase diagram methods, such as CALPHAD (Calculation of Phase Diagrams), are commonly employed to predict the stability of phases under various conditions [48].

2. Ab initio methods: Approaches like Density Functional Theory (DFT) are used to determine the pressures and temperatures at which phase transitions occur. These methods achieve this by assessing the differences in enthalpy between competing phases [49].

As an illustration, the phase stability of Ni-Ti alloys has been analyzed to elucidate their shape-memory characteristics and their ability to maintain stability at elevated temperatures [50].

6. Conclusion

This chapter provided an in-depth examination of the structural stability of materials, emphasizing the theoretical foundations and various stability aspects, including dynamical, mechanical, and thermodynamical stability. We highlighted the significance of several critical parameters—such as phonon dispersion, elastic constants, cohesive energy, and formation energy—as essential tools for assessing and predicting the stability of materials. Understanding how these parameters interact is fundamental to developing reliable materials for different applications.

6.1 Key takeaways

1. Theoretical insights: The combination of experimental data and first-principles computational techniques, such as Density Functional Theory (DFT), has greatly advanced our understanding of material stability. DFT, in particular, allows us to calculate key properties like energy differences between phases and transition states, providing crucial insights into how materials behave under different conditions. This synergy between theory and experiment has led to a deeper and more accurate understanding of the underlying principles of material stability.

2. Design for stability: Advances in materials design have been propelled by powerful computational tools, such as CALPHAD (Calculation of Phase Diagrams) and high-throughput screening methods. These tools enable the systematic design of materials with tailored stability properties. By simulating various environmental conditions and assessing stability across different compositions, researchers can predict and create materials that are specifically engineered to withstand certain conditions, such as high temperatures, extreme pressures, or corrosion. This approach is critical for developing materials for specialized applications, such as aerospace or energy technologies.

3. Challenges ahead: Despite the significant progress in understanding and designing stable materials, challenges remain. One of the major difficulties is predicting metastable materials, which may exist in non-equilibrium states but still exhibit useful properties. Additionally, ensuring material stability in extreme environments—such as in high-radiation or high-temperature conditions—poses a considerable challenge. Addressing these issues requires more sophisticated experimental methods and computational techniques that can model the complexities of real-world conditions.

Future Directions: Future research should focus on the integration of multiscale simulations, machine learning algorithms, and advanced experimental techniques to enhance the accuracy and efficiency of material stability predictions. Multiscale simulations, for example, can bridge different levels of analysis, from atomic-level interactions to bulk material properties, offering a more comprehensive understanding of stability. Machine learning can accelerate this process by identifying patterns and predicting material behaviors faster and with greater precision. These future efforts will be crucial for overcoming current limitations and addressing the growing need for advanced materials in fields like sustainable energy, high-performance electronics, and applications requiring resilience under extreme conditions. By improving our understanding of material stability, we can design better, more reliable materials that meet the demands of next-generation technologies and contribute to global challenges, such as energy sustainability and environmental resilience.

Acknowledgements

The author acknowledges the limited use of ChatGPT for language polishing of the chapter.

Conflict of interest

The authors declare no conflict of interest.

Author details

Amir Aliakbari* and Peiman Amiri
Department of Physics, Faculty of Science, Shahid Chamran University of Ahvaz, Ahvaz, Iran

*Address all correspondence to: amiraliakbari1369@gmail.com

References

[1] Born M, Huang K. Dynamical Theory of Crystal Lattices. England: Oxford University Press; 1996. DOI: 10.1093/oso/9780192670083.001.0001

[2] Geim AK, Novoselov KS. The rise of graphene. Nature Materials. 2007;**6**(3):183-191. DOI: 10.1038/nmat1849

[3] deWit R. Elastic constants and thermal expansion averages of a nontextured polycrystal. Journal of Mechanics of Materials and Structures. 2008;**3**(2):195-212. DOI: 10.2140/jomms.2008.3.195

[4] Monari A, Rivail JL, Assfeld X. Theoretical modeling of large molecular systems. Advances in the local self consistent field method for mixed quantum mechanics/molecular mechanics calculations. Accounts of Chemical Research. 2013;**46**(2):596-603. DOI: 10.1021/ar300278j

[5] Togo A, Tanaka I. First principles phonon calculations in materials science. Scripta Materialia. 2015;**108**:1-5. DOI: 10.1016/j.scriptamat.2015.07.021

[6] Aliakbari A, Amiri P. Structural, elastic, electronic, thermal, and phononic properties of yttrium carbide: First-principles calculations. Materials Chemistry and Physics. 2021;**270**:124744. DOI: 10.1016/j.matchemphys.2021.124744

[7] Martin RM. Electronic Structure: Basic Theory and Practical Methods. 2nd ed. Netherlands: Cambridge University Press; 2020. DOI: 10.1017/9781108555586

[8] Mouhat F, Coudert FX. Necessary and sufficient elastic stability conditions in various crystal systems. Physical Review B. 2014;**90**(22):224104. DOI: 10.1103/PhysRevB.90.224104

[9] Kohn W, Sham LJ. Self-consistent equations including exchange and correlation effects. Physical Review. 1965;**140**(4A):A1133. DOI: 10.1103/PhysRev.140.A1133

[10] Shapiro NZ, Shapley LS. Mass action laws and the Gibbs free energy function. Journal of the Society for Industrial and Applied Mathematics. 1965;**13**(2):353-375. DOI: 10.1137/0113020

[11] Schwabl F. Statistical Mechanics. Germany: Springer Science & Business Media; 2006. DOI: 10.1007/3-540-36217-7

[12] Colmenero F, Fernández AM, Cobos J, Timón V. Temperature-dependent Gibbs free energies of reaction of uranyl-containing materials based on density functional theory. The Journal of Physical Chemistry C. 2018;**122**(10):5268-5279. DOI: 10.1021/acs.jpcc.7b12368

[13] Nye JF. Physical Properties of Crystals: Their Representation by Tensors and Matrices. England: Oxford University Press; 1985. DOI: 10.1088/0031-9112/36/12/027

[14] Aliakbari A, Amiri P, Salehi H. First-principles investigation of the structural and dynamical stability, electronic and thermal properties of two-dimensional Yn+ 1Cn (n= 1, 2, and 3) MXenes. FlatChem. 2022;**31**:100328. DOI: 10.1016/j.flatc.2021.100328

[15] Amoudeh Z, Amiri P, Aliakbari A. The new solid solution of double transition metal MXenes: Atomistic modeling of two-dimensional YScX (X= C and N). Solid State Sciences. 2023;**144**:107306. DOI: 10.1016/j.solidstatesciences.2023.107306

[16] Su H, Fang X, Feng X, Yan B. The temperature-dependent strength of

metals: Theory and experimental validation. Journal of Applied Mechanics. 2014;**81**(9):091003. DOI: 10.1115/1.4027814

[17] Burke K. Perspective on density functional theory. The Journal of Chemical Physics. 2012;**136**(15):1-10. DOI: 10.1063/1.4704546

[18] Freking U, Mazur A, Pollmann J. Surface phonons of S: Si (001)-(1× 1). Physical Review B. 2001;**64**(24):245341. DOI: 10.1103/PhysRevB.64.245341

[19] Baroni S, De Gironcoli S, Dal Corso A, Giannozzi P. Phonons and related crystal properties from density-functional perturbation theory. Reviews of Modern Physics. 2001;**73**(2):515. DOI: 10.1103/RevModPhys.73.515

[20] Zhong W, King-Smith RD, Vanderbilt D. Giant LO-TO splittings in perovskite ferroelectrics. Physical Review Letters. 1994;**72**(22):3618. DOI: 10.1103/PhysRevLett.72.3618

[21] Pop E, Varshney V, Roy AK. Thermal properties of graphene: Fundamentals and applications. MRS Bulletin. 2012;**37**(12):1273-1281. DOI: 10.1557/mrs.2012.203

[22] Hoffman AE, Temmerman W, Campbell E, Damin AA, Lezcano-Gonzalez I, Beale AM, et al. A critical assessment on calculating vibrational spectra in nanostructured materials. Journal of Chemical Theory and Computation. 2023;**20**(2):513-531. DOI: 10.1021/acs.jctc.3c00942

[23] Marronnier A, Lee H, Geffroy B, Even J, Bonnassieux Y, Roma G. Structural instabilities related to highly anharmonic phonons in halide perovskites. The Journal of Physical Chemistry Letters. 2017;**8**(12): 2659-2665. DOI: 10.1021/acs.jpclett.7b00807

[24] Bux SK, Blair RG, Gogna PK, Lee H, Chen G, Dresselhaus MS, et al. Nanostructured bulk silicon as an effective thermoelectric material. Advanced Functional Materials. 2009; **19**(15):2445-2452. DOI: 10.1002/adfm.200900250

[25] Pennelli G, Dimaggio E, Masci A. Silicon nanowires: A breakthrough for thermoelectric applications. Materials. 2021;**14**(18):5305. DOI: 10.3390/ma14185305

[26] Kortus J, Mazin II, Belashchenko KD, Antropov VP, Boyer LL. Superconductivity of metallic boron in MgB 2. Physical Review Letters. 2001;**86**(20):4656. DOI: 10.1103/PhysRevLett.86.4656

[27] Lindsay L, Broido DA, Reinecke TL. First-principles determination of ultrahigh thermal conductivity of boron arsenide: A competitor for diamond? Physical Review Letters. 2013;**111**(2): 025901. DOI: 10.1103/PhysRevLett.111.025901

[28] Morin FJ. Oxides which show a metal-to-insulator transition at the Neel temperature. Physical Review Letters. 1959;**3**(1):34. DOI: 10.1103/PhysRevLett.3.34

[29] Hu P, Hu P, Vu TD, Li M, Wang S, Ke Y, et al. Vanadium oxide: Phase diagrams, structures, synthesis, and applications. Chemical Reviews. 2023; **123**(8):4353-4415. DOI: 10.1021/acs.chemrev.2c00546

[30] Amiri P, Aliakbari A, Behzadi P, Ketabi SA. The stability, electronic, thermal, and optical properties of silver halide (AgX: X= F, Cl, Br, and I) semiconductors: Ab-initio study.

Computational Condensed Matter. 2023; **37**:e00837. DOI: 10.1016/j.cocom.2023.e00837

[31] Field JE. The mechanical and strength properties of diamond. Reports on Progress in Physics. 2012;**75**(12): 126505. DOI: 10.1088/0034-4885/75/12/126505

[32] Amoudeh Z, Amiri P, Aliakbari A. Ab-initio investigation of stability and electronic properties of new yttrium-based solid solution of double transition metal YMXTx (M= Ti and zr; X= C and N; Tx= H, O, and F) MXenes. Physica B: Condensed Matter. 2024;**691**:416363. DOI: 10.1016/j.physb.2024.416363

[33] Aliakbari A, Amiri P, Dezfuli AG. Stability and physical properties of yttrium-based new MAX phases Y_2AX (a= Al, Si, Ga, and Ge; X= C and N): A first-principles prediction. Applied Physics A. 2023;**129**(7):476. DOI: 10.1007/s00339-023-06743-6

[34] Born M, Huang K. Dynamical Theory of Crystal Lattices. England: Oxford University Press; 1996. DOI: 10.1119/1.1934059

[35] Wu ZJ, Zhao EJ, Xiang HP, Hao XF, Liu XJ, Meng J. Crystal structures and elastic properties of superhard IrN2 and IrN3 from first principles. Physical Review B—Condensed Matter and Materials Physics. 2007;**76**(5):054115. DOI: 10.1103/PhysRevB.76.054115

[36] Singh S, Wu Q, Yue C, Romero AH, Soluyanov AA. Topological phonons and thermoelectricity in triple-point metals. Physical Review Materials. 2018;**2**(11): 114204. DOI: 10.1103/PhysRevMaterials.2.114204

[37] Maździarz M. Comment on 'the computational 2D materials database: High-throughput modeling and discovery of atomically thin crystals'. 2D Materials. 2019;**6**(4):048001. DOI: 10.1088/2053-1583/ab2ef3

[38] Amiri P, Mokhtaripoor N, Aliakbari A, Salehi H. Ab-initio study of structural stability, electronic and optical properties of X2GeSe (X= P and Sb) monolayers. Solid State Communications. 2022;**343**:114673. DOI: 10.1016/j.ssc.2022.114673

[39] Zebarjad S, Amiri P, Aliakbari A, Salehi H. Ab-initio investigation of the structural stability, electronic and optical properties of the LiBO2 compound by using the G0W0+ BSE approach. Computational Condensed Matter. 2023; **34**:e00789. DOI: 10.1016/j.cocom.2023.e00789

[40] Abeshipour A, Amiri P, Aliakbari A, Salehi H. An investigation of dynamical stability, electronic, and optical properties of sodium metaborate (NaBO2): A many-body G0W0+ BSE approach. Solid State Communications. 2023;**360**:115055. DOI: 10.1016/j.ssc.2022.115055

[41] Li XH, Su XY, Zhang RZ, Xing CH, Zhu ZL. Pressure-induced band engineering, work function and optical properties of surface F-functionalized Sc2C MXene. Journal of Physics and Chemistry of Solids. 2020;**137**:109218. DOI: 10.1016/j.jpcs.2019.109218

[42] Shin H, Kang S, Koo J, Lee H, Kim J, Kwon Y. Cohesion energetics of carbon allotropes: Quantum Monte Carlo study. The Journal of Chemical Physics. 2014; **140**(11):114702. DOI: 10.1063/1.4867544

[43] Kirklin S, Saal JE, Meredig B, Thompson A, Doak JW, Aykol M, et al. The open quantum materials database (OQMD): Assessing the accuracy of DFT formation energies. npj Computational

Materials. 2015;**1**(1):1-5. DOI: 10.1038/npjcompumats.2015.10

[44] Amiri P, Shavi ZD, Aliakbari A, Salehi H. The investigation of the half-metallic properties of half-Heusler KXP (X= Cr & Mo) compounds: A first-principles study. Journal of Physics and Chemistry of Solids. 2022;**160**:110294. DOI: 10.1016/j.jpcs.2021.110294

[45] Diebold U. The surface science of titanium dioxide. Surface Science Reports. 2003;**48**(5–8):53-229. DOI: 10.1016/S0167-5729(02)00100-0

[46] Shah SH, Bristowe PD. Point defect formation in M2AlC (M= Zr, Cr) MAX phases and their tendency to disorder and amorphize. Scientific Reports. 2017;**7**(1):9667. DOI: 10.1038/s41598-017-10273-6

[47] Zhang B, Tan R, Yang L, Zheng J, Zhang K, Mo S, et al. Mechanisms and properties of ion-transport in inorganic solid electrolytes. Energy Storage Materials. 2018;**10**:139-159. DOI: 10.1016/j.ensm.2017.08.015

[48] Luo Q, Guo Y, Liu B, Feng Y, Zhang J, Li Q, et al. Thermodynamics and kinetics of phase transformation in rare earth–magnesium alloys: A critical review. Journal of Materials Science & Technology. 2020;**44**:171-190. DOI: 10.1016/j.jmst.2020.01.022

[49] da Silva EL, Santos MC, Rodríguez-Hernández P, Muñoz A, Manjón FJ. Theoretical study of pressure-induced phase transitions in Sb2S3, Bi2S3, and Sb2Se3. Crystals. 2023;**13**(3):498. DOI: 10.3390/cryst13030498

[50] Ma J, Karaman I, Noebe RD. High temperature shape memory alloys. International Materials Reviews. 2010;**55**(5):257-315. DOI: 10.1179/095066010X12646898728363

Chapter 4

Holographic-Type Behaviors in Complex Systems Dynamics from a Multifractal Perspective of Motion

Catalin Gabriel Dumitras, Vlad Ghizdovat, Cristina Maricela Rusu, Emanuel Nazaretian, Florin Nedeff, Valentin Nedeff, Iosif Ioja and Maricel Agop

Abstract

A unitary model of complex systems dynamics is proposed, assuming that these systems can be assimilated to a multifractal mathematical object. Then, if we describe these dynamics in the framework of The Scale Relativity Theory, through continuous and non-differentiable curves (fractal or multifractal curves), possible holographic-type behaviors can be obtained. In this context, the Schrödinger and Madelung multifractal scenarios align through the generation of patterns, symmetries of the normalized velocity relative to the symmetry axis of the spatial-temporal Gaussian, and vertices at the periphery of the pattern for the normalized velocity field (Taylor-type effect). In conclusion, we propose a unitary model for describing complex systems dynamics, on both monofractal and multifractal manifolds.

Keywords: multifractal, complex system, dynamics, shape memory alloy, holographic-type behavior

1. Introduction

Complex systems are extensive multidisciplinary fields of study investigated by a mixture of basic theories, particularly from physics and computational simulations. This system has several interacting elemental components referred to as "agents." Complex systems encompass ecosystems, characterized by organism-environment interactions; the human brain, featuring intricate neural networks; the Internet, linking millions of devices and users; financial markets, marked by dynamic economic interactions; and weather systems, comprising numerous variables that affect climate patterns and forecasting [1–4].

The manifestation of such a system cannot be anticipated only by the behavior of individual components or by aggregating their behaviors. Rather, it is dictated by the way in which components interact to affect global behavior. Key characteristics of complex systems include emergence, self-organization, and adaptability, among others [5, 6].

Polymers are a prime example of a complex system. Their structures exhibit a range of organizations, from basic, linear sequences of identical structural units to intricate chains composed of amino acid sequences that constitute the fundamental components of living organisms. One of the most fascinating complex systems in nature is DNA, which encodes the information that governs all biological activities. It is accountable for the extraordinary manner in which individual cells arrange into intricate systems like organs, which subsequently constitute even more complicated systems like organisms. The examination of complex systems can provide insight into the realistic dynamics of polymers, addressing challenging issues such as protein folding [1–4].

Consequently, the theoretical models that elucidate complex systems dynamics grow increasingly sophisticated [1–4]. Nonetheless, the situation can be standardized, considering that the intricacy of the interaction process necessitates multiple temporal resolution scales, and the evolution of patterns imposes differing degrees of freedom [7].

To formulate new theoretical models, we must take into account that complex systems demonstrate chaotic behavior *via* self-similarity, characterized by the repetition of patterns at various space-time scales. In systems such as meteorology or financial markets, little alterations can result in significant, unforeseen consequences. This fractal repeating at several layers gives a sense of order among perceived chaos, enhancing the system's complexity [1–5]. For extensive temporal scales relative to the inverse of the largest Lyapunov exponent, deterministic trajectories are supplanted by a set of possible trajectories, and the notion of precise positions is substituted with that of probability density. In complex systems, collision processes may be analyzed using particle dynamics on non-differentiable curves, indicating that particle trajectories are irregular and have fractal characteristics. The non-smooth trajectories indicate that little alterations in initial conditions result in markedly different results, increasing system sensitivity. Particle collisions frequently lead to abrupt, nonlinear alterations in their states, generating feedback loops. The presence of these non-differentiable curves indicates that minor disturbances may escalate, leading to chaotic behavior inside the system. The irregular dynamics hinder prediction due to the system's development being extremely sensitive to beginning circumstances and microscopic interactions.

Given that non-differentiability is a universal characteristic of complex systems, it is essential to develop a non-differentiable physics. In this hypothesis, by positing that the complexity of interaction processes is supplanted by non-differentiability, it becomes unnecessary to employ the entire classical repertoire of quantities from conventional physics (differentiable physics).

2. Hallmarks of fractality

Let us quickly remember what fractality means [8]:

- Utilizing nonlinear dynamics and chaos theory to analyze phenomena inadequately addressed by classical mechanics, for a description grounded in mathematical modeling, employing formulas tailored to the continuous fluctuations arising from intricate physical phenomena; we will delineate and accept a tangible component and a "virtual" one (which can be activated contingent upon specific conditions at any given moment);

- A nonlinear object appears distinct based on the resolution scale of observation; hence, the complexity profile is correlated with the resolution scale employed during observation.

- Sensitivity to initial conditions: In predictable systems, a minor alteration in the initial state results in a correspondingly minor change in the final state; however, in chaotic systems, especially biological ones, small perturbations can result in divergent and unpredictable outcomes, with the "butterfly effect" serving as a prominent illustration of this phenomenon. In 1963, Edward Lorentz became the inaugural meteorologist to provide a rational elucidation for the inherent probabilistic nature of meteorological forecasts: He posited, a notion subsequently corroborated by other scholars, that the flapping of a butterfly's wings in Europe could precipitate a tornado in America under specific, largely unpredictable conditions, which defy the precision of meteorological models;

- The applicability of the superposition principle differs: In a linear system, the resultant effect of two distinct causes is the aggregate of the individual effects of each cause; conversely, in a nonlinear system, the combination of two elementary actions may yield novel effects due to interactions among constituent elements, resulting in the emergence of structures and events that are unpredictable in both space and time—deterministic chaos.

Benoît B. Mandelbrot, the mathematician who popularized the notion of fractals, delineated six essential characteristics that constitute a fractal as a geometric entity. Mandelbrot posits that fractals possess the following attributes [8]:

- Self-similarity: Fractals frequently display self-similarity, indicating that they reveal analogous patterns or forms at different sizes. This self-similarity may be precise (same structure across varying scales) or statistical (similar patterns with minor differences across different sizes). The Mandelbrot set, a renowned fractal, demonstrates self-similarity in its complex boundary structures upon magnification.

- Fractional (non-integer) dimension: In contrast to conventional geometric forms such as lines, planes, or cubes, which possess integer dimensions (1D, 2D, or 3D), fractals frequently exhibit a fractal dimension that is non-integer. This indicates that they exist in a continuum between integer dimensions. Mandelbrot established the notion of fractal dimension to measure the complexity of these entities. A fractal dimension often resides between the dimensions of a curve (1D) and a surface (2D), signifying the complexity of the structure;

- Fractals exhibit complexity at all scales. Regardless of the degree of magnification, the fractal maintains its elaborate, complicated design. The recursive complexity is a fundamental characteristic that differentiates fractals from conventional geometric entities, which often exhibit smoothness upon magnification;

- Fractals exhibit limitless detail with sufficient magnification. As one examines ever smaller sizes, further features arise, uncovering an infinite complexity of structure. In the Mandelbrot set, magnifying its boundaries uncovers increasing complexity, and regardless of the extent of magnification, a point of simple resolution is never attained;

- Fractals are often formed using an iterative or recursive process. The Mandelbrot set is generated by the repetitive iteration of a mathematical algorithm. The process produces progressively intricate structures with each iteration, and the set's boundary reflects this recursive characteristic;
- Fractals frequently display irregularity or roughness. In contrast to smooth curves or surfaces, fractals lack basic, regular geometric forms such as circles or squares. The irregularity of fractals contributes to their "wild" esthetic;
- Scale invariance: Numerous fractals demonstrate scale invariance, indicating that their structure seems consistent irrespective of the magnification level. This characteristic renders fractals more intriguing in disciplines such as physics, where scale invariance is evident in natural occurrences like coastlines or mountain ranges.

The fundamental concepts of fractal geometry include:

- Initiator: The geometric figure from which the fractal originates, typically a basic geometric shape such as a line, square, circle, or rhombus;
- Construction law: The procedure employed to generate the fractal;
- Generating process: The mechanism that systematically creates iterations of the fractal object, commencing from the current iteration and applying the construction law, whereby the repetition of this step produces new generations of fractal sets.

Fractals may be employed to construct models of uncertain and chaotic processes.

Let us now assume that the motion of the entities inside complex systems occurs along fractal curves, which are continuous but non-differentiable. A manifold that facilitates such movement defines a fractal space-time. The fractal characteristics of space-time suggest that non-differentiability leads to the breaking of differential time reflection invariance. In this context, the standard definitions of the derivative of a certain function with respect to time [9–14]:

$$\frac{df}{dt} = \lim_{\Delta t \to 0} \frac{f(t + \Delta t) - f(t)}{\Delta t} = \lim_{\Delta t \to 0} \frac{f(t) - f(t - \Delta t)}{\Delta t} \tag{1}$$

are, in the differentiable case, equivalent. Transition occurs through transformation $\Delta t \to -\Delta$ (time reflection invariance at the infinitesimal level). In the non-differentiable case, two functions (df_{+}/dt) and (df_{-}/dt) are defined as explicit functions of t and dt:

$$\begin{aligned} \frac{df_{+}}{dt} &= \lim_{\Delta t \to 0} \frac{f(t + \Delta t, \Delta t) - f(t, \Delta t)}{\Delta t} \\ \frac{df_{-}}{dt} &= \lim_{\Delta t \to 0} \frac{f(t, \Delta t) - f(t - \Delta t, \Delta t)}{\Delta t} \end{aligned} \tag{2}$$

where "+" corresponds to forward processes and "−" to the backward ones. Then, in spatial coordinates dX, we can write [9–11]:

$$dX_\pm = dx_\pm + d\zeta_\pm = v_\pm dt + d\zeta_\pm \tag{3}$$

with $v_\pm$ the forward and backward mean velocities:

$$v_+ = \frac{dx_+}{dt} = \lim_{\Delta t \to 0_+} \left(\frac{X(t+\Delta t) - X(t)}{\Delta t} \right)$$
$$v_- = \frac{dx_-}{dt} = \lim_{\Delta t \to 0_-} \frac{(X(t) - X(t-\Delta t)}{\Delta t} \tag{4}$$

and $d\xi_\pm$ a measure of non-differentiability (a variation caused by the fractal characteristics of the trajectory) with the mean:

$$\langle d\xi_\pm \rangle = 0 \tag{5}$$

Although the concept of velocity is traditionally singular, if space-time is fractal, it necessitates the introduction of two velocities (v_+ and v_-). The "two-values" of the velocity vector is a novel, unique outcome of non-differentiability that lacks a conventional counterpart in differential physics.

However, we cannot prefer v_+ over v_-. The only solution is to consider both the forward ($dt > 0$) and backward ($dt < 0$) processes together. In this context, it is mandatory to introduce the complex velocity [9–11]:

$$V = \frac{v_+ + v_-}{2} - i\frac{v_+ - v_-}{2} = \frac{dx_+ + dx_-}{2dt} - i\frac{dx_+ - dx_-}{2dt}, i = \sqrt{-1} \tag{6}$$

If $(v_+ + v_-)/2$ may be considered as differentiable (classical) velocity, then the difference $(v_+ - v_-)/2$ is the non-differentiable (fractal) velocity.

Implementing the notations $dx_\pm = d_\pm x$, Eq. (6) becomes:

$$V = \left(\frac{d_+ + d_-}{2dt} - i\frac{d_+ - d_-}{2dt} \right) x \tag{7}$$

Thus, we can define the operator:

$$\frac{\hat{d}}{dt} = \frac{d_+ + d_-}{2dt} - i\frac{d_+ - d_-}{2dt} \tag{8}$$

In this conjecture, let us assume that the fractal curve is situated within a three-dimensional space and that $\boldsymbol{X}$ of components X^i (i = $\overline{1,3}$) is the position vector of a point on the curve. Let us also consider a function $f(X,t)$ and the following Taylor series expansion up to the second order:

$$df = f(X^i + dX^i, t + dt) - f(X^i, dt) = \left(\frac{\partial}{\partial X^i} dX^i + \frac{\partial}{\partial t} dt \right) f(X^i, t) + \frac{1}{2} \left(\frac{\partial}{\partial X^i} dX^i + \frac{\partial}{\partial t} dt \right)^2 f(X^i, t) \tag{9}$$

From here, using the notations, $dX_\pm{}^i = d_\pm X^i$, the forward and backward average values of this relation take the form:

$$\langle d_\pm f\rangle = \langle\frac{\partial f}{\partial t}dt\rangle + \langle\nabla f\cdot d_\pm X\rangle + \frac{1}{2}\langle\frac{\partial^2 f}{\partial t^2}(dt)^2\rangle + \langle\frac{\partial^2 f}{\partial X^i\partial t}d_\pm X^i dt\rangle + \frac{1}{2}\langle\frac{\partial^2 f}{\partial X^i\partial X^l}d_\pm X^i d_\pm X^l\rangle \tag{10}$$

We establish the following conditions: the mean values of the function f and its derivatives coincide with each other, and the differentials $d_\pm X^i$ and dt are independent; therefore, the averages of their products coincide with the product of average. Thus, Eq. (10) becomes:

$$d_\pm f = \frac{\partial f}{\partial t}dt + \nabla f\langle d_\pm X\rangle + \frac{1}{2}\frac{\partial^2 f}{\partial t^2}\langle(dt)^2\rangle + \frac{\partial^2 f}{\partial X^i\partial t}\langle d_\pm X^i dt\rangle + \frac{1}{2}\frac{\partial^2 f}{\partial X^i\partial X^l}\langle d_\pm X^i d_\pm X^l\rangle$$

or more, by using Eq. (3):

$$d_\pm f = \frac{\partial f}{\partial t}dt + \nabla f d_\pm x + \frac{1}{2}\frac{\partial^2 f}{\partial t^2}(dt)^2 + \frac{\partial^2 f}{\partial X^i\partial t}d_\pm x^i dt + \frac{1}{2}\frac{\partial^2 f}{\partial X^i\partial X^l}\left(d_\pm x^i d_\pm x^l + \langle d\xi^i_\pm d\xi^l_\pm\rangle\right) \tag{11}$$

Since $d\xi^i_\pm$ describes the fractal properties of the trajectory with the fractal dimension D_F [8], it is natural to impose $\left(d\xi^i_\pm\right)^{D_F}$ to be proportional with dt, that is, [9–11]:

$$\left(d\xi^i_\pm\right)^{D_F} = \sqrt{2D}dt \tag{12}$$

where D is a coefficient of proportionality (the fractal non-fractal transition coefficient).

Let us now focus on the mean $\langle d\xi^i_\pm d\xi^l_\pm\rangle$. If $i \neq l$, this average is zero due to the independence of $d\xi^i$ and $d\xi^l$. So, using Eq. (13), we can write:

$$\langle d\xi^i_\pm d\xi^l_\pm\rangle = \pm\delta^{il}2D(dt)^{\left(\frac{2}{D_F}\right)-1} \tag{13}$$

with:

$$\delta^{il} = \begin{cases} 1, if i = l \\ 0, if i \neq l \end{cases}$$

and we had considered that:

$$\begin{cases} \langle d\xi^i_+ d\xi^l_+\rangle > 0 \; and \; dt > 0 \\ \langle d\xi^i_- d\xi^l_-\rangle > 0 \; and \; dt < 0 \end{cases} \tag{14}$$

Then, Eq. (12) may be written under the form:

$$d_\pm f = \frac{\partial f}{\partial t}dt + \nabla f d_\pm x + \frac{1}{2}\frac{\partial^2 f}{\partial t^2}(dt)^2 + \frac{\partial^2 f}{\partial X^i\partial t}d_\pm x^i dt + \frac{\partial^2 f}{\partial X^i\partial X^l}d_\pm x^i d_\pm x^l \pm \frac{\partial^2 f}{\partial X^i\partial X^l}\delta^{il}D(dt)^{(2/D_F)-1} \tag{15}$$

If we divide by dt and neglect the terms that contain differential factors, Eq. (15) is reduced to:

$$\frac{d_{\pm}f}{dt} = \frac{\partial f}{\partial t} + v_{\pm}\nabla f_{\pm} \pm D(dt)^{\left(\frac{2}{D_F}\right)-1}\Delta f \tag{16}$$

In this context, we will calculate $\frac{\hat{d}f}{dt}$. According to Eq. (8) and considering Eq. (16), we have:

$$\begin{aligned}\frac{\hat{d}f}{dt} &= \frac{1}{2}\left[\frac{d_+f}{dt} + \frac{d_-f}{dt} - i\left(\frac{d_+f}{dt} - \frac{d_-f}{dt}\right)\right] = \frac{1}{2}\left[\left(\frac{\partial f}{\partial t} + v_+\nabla f + D(dt)^{(2/D_F)-1}\Delta f\right) + \left(\frac{\partial f}{\partial t} + v_-\nabla f - D(dt)^{(2/D_F)-1}\Delta f\right)\right]\\ &-\frac{i}{2}\left[\left(\frac{\partial f}{\partial t} + v_+\nabla f + D(dt)^{(2/D_F)-1}\Delta f\right) - \left(\frac{\partial f}{\partial t} + v_-\nabla f - D(dt)^{(2/D_F)-1}\Delta f\right)\right]\\ &= \frac{\partial f}{\partial t} + \left(\frac{v_+ + v_-}{2} - i\frac{v_+ - v_-}{2}\right)\nabla f - iD(dt)^{\left(\frac{2}{D_F}\right)-1}\Delta f\end{aligned} \tag{17}$$

or, by using Eq. (6):

$$\frac{\hat{d}f}{dt} = \frac{\partial f}{\partial t} + \boldsymbol{V}\cdot\nabla f - iD(dt)^{\left(\frac{2}{D_F}\right)-1}\Delta f \tag{18}$$

This relation also allows us to give the definition of the fractal operator:

$$\frac{\hat{d}}{dt} = \frac{\partial}{\partial t} + \boldsymbol{V}\cdot\nabla - iD(dt)^{(2/D_F)-1}\Delta \tag{19}$$

By applying the fractal operator (Eq. (19)) to the complex velocity (Eq. (6)) and by accepting the principle of scale covariance [9, 11] in the form:

$$\frac{\hat{d}\boldsymbol{V}}{dt} = -\nabla U \tag{20}$$

we can obtain the motion equation:

$$\frac{\hat{d}\boldsymbol{V}}{dt} = \frac{\partial \boldsymbol{V}}{\partial t} + (\boldsymbol{V}\cdot\nabla)\boldsymbol{V} - iDdt^{\left(\frac{2}{D_F}\right)-1}\Delta\boldsymbol{V} = -\nabla U \tag{21}$$

where U is an external scalar potential. Eq. (21) is a Navier-Stokes type equation. It signifies that at every juncture of a fractal trajectory, the local acceleration, $\partial_t\boldsymbol{V}$, the non-linearly (convective) term, $(\boldsymbol{V}\cdot\nabla)\boldsymbol{V}$, the dissipative term, $Ddt^{\left(\frac{2}{D_F}\right)-1}\Delta\boldsymbol{V}$, and the external free term ∇U make their balance. In this context, the complex fluid can be assimilated to a "rheological" fluid, whose dynamics are described by the complex velocities field, $\boldsymbol{V}$, and by the imaginary viscosity type coefficient, $iDdt^{\left(\frac{2}{D_F}\right)-1}$. The fluid's rheology can confer hysteretic characteristics to the complex fluid, which exhibits a hysteresis cycle and memory [15–19].

3. Chaoticity through turbulence and stochasticization *via* fractality

For irrotational motions of the complex systems' entities:

$$\nabla\times\boldsymbol{V} = 0, \quad \nabla\times\boldsymbol{V}_D = 0, \quad \nabla\times\boldsymbol{V}_F = 0 \tag{22}$$

$\boldsymbol{V}$ can be chosen as:

$$\hat{\boldsymbol{V}} = -2iDdt^{\left(\frac{2}{D_F}\right)-1}\nabla\, ln\, \psi \tag{23}$$

where $\phi \equiv ln\, \psi$ is the velocity scalar potential. By substituting Eq. (23) in Eq. (21) and using the method described in Refs. [15–19], it results:

$$\frac{d\hat{\boldsymbol{V}}}{dt} = -iDdt^{\left(\frac{2}{D_F}\right)-1}\nabla\left[\frac{\partial\, ln\, \psi}{\partial t} - iDdt^{\left(\frac{2}{D_F}\right)-1}\frac{\nabla\psi}{\psi} + U\right] = 0 \tag{24}$$

This equation can be integrated in a universal way and yields:

$$D^2 dt^{\left(\frac{4}{D_F}\right)-2}\Delta\psi + iDdt^{\left(\frac{2}{D_F}\right)-1}\frac{\partial\psi}{\partial t} - \frac{U}{2}\psi = 0 \tag{25}$$

up to an arbitrary phase factor that can be nullified by an appropriate selection of the phase of ψ. Eq. (25) is a Schrödinger-type equation—we will call it the Schrödinger scenario. For motions on Peano curves, $D_F = 2$ [8] at Compton scale $D = \ /2m_0$ [9, 11], with $_0$ the reduced Planck constant and m_0 the rest mass of the complex system's entities, Eq. (25) becomes the standard Schrödinger equation [20]:

$$\frac{\hbar^2}{2m_0}\Delta\psi + i\hbar\frac{\partial\psi}{\partial t} - \frac{U}{2m_0}\psi = 0 \tag{26}$$

If $\psi = \sqrt{\rho}e^{iS}$, with $\sqrt{\rho}$ the amplitude and S the phase of ψ, the complex velocity field (Eq. (6)) takes the explicit form:

$$\boldsymbol{V} = 2Ddt^{\left(\frac{2}{D_F}\right)-1}\nabla S - iDdt^{\left(\frac{2}{D_F}\right)-1}\nabla\, ln\, \rho \tag{27}$$

$$\boldsymbol{V}_D = 2Ddt^{\left(\frac{2}{D_F}\right)-1}\nabla S \tag{28}$$

$$\boldsymbol{V}_F = Ddt^{\left(\frac{2}{D_F}\right)-1}\nabla\, ln\, \rho \tag{29}$$

By substituting Eq. (27)–(29) in Eq. (21) and separating the real and imaginary components, subject to an arbitrary phase factor that can be nullified by an appropriate selection of the phase of ψ, we obtain:

$$\frac{\partial\boldsymbol{V}_D}{\partial t} + (\boldsymbol{V}_D\cdot\nabla)\boldsymbol{V}_D = -\nabla(Q+U) \tag{30}$$

$$\frac{\partial\rho}{\partial t} + \nabla\cdot(\rho\boldsymbol{V}_D) = 0 \tag{31}$$

with Q the specific fractal potential:

$$Q = -2Ddt^{\left(\frac{4}{D_F}\right)-2}\frac{\Delta\sqrt{\rho}}{\sqrt{\rho}} = -\frac{\boldsymbol{V}_F^2}{2} - Ddt^{\left(\frac{2}{D_F}\right)-1}\nabla\cdot\boldsymbol{V}_F \tag{32}$$

Eq. (30) represents the specific momentum conservation law, while Eq. (31) represents the states density conservation law. Through the fractal velocity, $\boldsymbol{V}_F$, the specific fractal potential Q quantifies the non-differentiability of the complex system's entities trajectories, reflecting their chaotic nature. Eqs. (30)–(32) define the fractal hydrodynamics model—we will call it the Madelung scenario. In such a context, the complex system is assimilated to a fractal fluid.

Now, certain conclusions are apparent:

i. Any particle is in a permanent interaction with the "sub-fractal level" through the specific fractal potential, Q;

ii. The "sub-fractal level" (fractal medium) is identified with a non-relativistic fractal fluid described by the states density and the momentum conservation laws—see Eqs. (30) and (31). These equations correspond to the quantum hydrodynamics model. Indeed, for motions on fractal curves in the fractal dimension $D_F = 2$ at Compton scale, $D = /2m_0$ with the radius Planck constant, the fractal hydrodynamics model reduces to the quantum hydrodynamics model. Moreover, at Compton scale, the "sub-fractal level" can be identified with a "sub-quantum level" [21];

iii. The fractal potential arises from non-differentiability and should be seen as a kinetic term rather than a potential one. The fractal potential (32) can generate a type of viscosity stress tensor [18]:

$$\hat{\sigma}_{il} = m_0 D^2 (dt)^{\left(\frac{4}{D_F}\right)-2}\left[\nabla_i\rho\nabla_l\rho - \left(\frac{\nabla_i\rho\nabla_l\rho}{\rho}\right)\right] == \eta\left(\frac{\partial u_i}{\partial x_l}+\frac{\partial u_l}{\partial x_i}\right), \eta = \frac{1}{2}m_0\rho D(dt)^{\left(\frac{2}{D_F}\right)-1} \tag{33}$$

of which divergence is equal to the usual force density associated with Q:

$$\nabla_i\hat{\sigma}_{il} = -\rho\nabla_l Q \tag{34}$$

i. For motions on Peano curves at Compton scale [8–11], the fractal hydrodynamics model reduces to a quantum hydrodynamic model. The fractal velocity $\boldsymbol{V}_F$ does not denote actual mechanical motion; nonetheless, it does facilitate the transfer of specific momentum and the concentration of energy. The lack of $\boldsymbol{V}_F$ in the state's density conservation law and its function in the variational principle clearly illustrate this point. Any interpretation of Q must acknowledge the "self" or intrinsic quality of the specific momentum transfer, although energy is conserved in total, being stored as mass motion and potential energy in a classical sense, with some energy being accessible elsewhere. The conservation of energy and particular momentum guarantees reversibility and the presence of eigenstates while precluding a Brownian-motion kind of interaction with an external medium;

ii. For movements along Peano curves, at spatial scales higher than the boundary layer dimension and at temporal scales surpassing the oscillation periods of the pulsating velocities that coincide with the average velocity of the complex fluid motions (for details see Refs. [22, 23]) the fractal hydrodynamics model reduces to the standard hydrodynamical model [24]. We must highlight the fact that, in this approximation of motion and for a certain type of complex fluid [2, 5], $\nabla\bar{\sigma} = -\nabla p/\rho$, with p the pressure;

iii. Since the position vector of the complex fluid entity can be associated with a Wiener-type stochastic process of [8–11], ψ is not only the scalar potential of a complex velocity (through $\phi \equiv ln\,\psi$) in the frame of the fractal hydrodynamics model, but it also represents states density (through $|\psi|^2$) in the frame of a Schrödinger-type model. It results in the equivalence between the two formalisms (FHM and the Schrödinger-type one). Furthermore, chaotic behavior, whether arising from turbulence in the fractal hydrodynamics framework or from stochastic processes in the Schrödinger-type framework, is only produced by the non-differentiability of movement trajectories inside a fractal space;

iv. In the standard model (Landau's scenario [24, 25]), the Fourier spectrum is invariably discrete and cannot approximate a continuous spectrum except in instances involving a substantial number of frequencies, which produce an infinite number of spectral components due to the beats arising from nonlinearities within the complex system. However, according to the conventional model, the flow can never be genuinely chaotic because, in the presence of many periodic functions, correlations are not null but have an oscillatory nature. Consequently, Landau's scenario may elucidate the transition to chaotic behavior alone in a complex fluid possessing an unlimited number of degrees of freedom. In our case, when $\delta t/\tau \to 0$ for $D_F \neq 2$ the physical parameters that characterize the behavior of the complex system are no longer delineated. In this approximation, a simulation of a system possessing an unlimited number of degrees of freedom is employed. Furthermore, dynamic states may emerge, defined by intervals of regular oscillations disrupted by chaotic surges, with the transition between these states occurring spontaneously, unpredictably, and independently of any variations in control settings (turbulence by intermittency);

v. The function $\psi(\boldsymbol{r}, t)$ is invariant with respect to a change of its phase $S(\boldsymbol{r}, t)$ by an integer multiple of 2π. Hence, Eq. (34) gives the circulation [20]:

$$\Gamma_D = \oint m_0 \boldsymbol{V}_D d\boldsymbol{r} = 2D\oint dS = 4\pi m_0\, \mathrm{Dm},\ \mathrm{m} = 0, 1, \pm 2 \qquad (35)$$

where D is scale dependent. In particular, for movements on Peano-type fractal curves with $D_F = 2$ (congruent with Brownian motion), at Compton scale, $D = \frac{\ }{2m_0}$, Eq. (35) takes the standard form:

$$\Gamma_D = \oint m_0 \boldsymbol{V}_D d\boldsymbol{r} = mh \qquad (36)$$

The path of integration of Eq. (35) may not pass through regions where $\rho = 0$, that is, where $\boldsymbol{V}_D$ is potentially singular but is otherwise arbitrary.

Eqs. (30) and (31) represent a complete system of differential equations for the field $\rho(r, t)$ and $\boldsymbol{V}_D(\boldsymbol{r}, t)$. Eq. (36) relates each non-differentiable hydrodynamic solution $(\rho, \boldsymbol{V}_D, m)$ to a solution ψ in a unique way.

The field $\rho(\boldsymbol{r}, t)$ has the meaning of a probability distribution; that is, the probability of finding the particle in the vicinity $d\boldsymbol{r}$ of the point $\boldsymbol{r}$, is, at the time t:

$$dP = \rho d\boldsymbol{r}, \iiint \rho d\boldsymbol{r} = 1 \tag{37}$$

Spatial integrations extend over the entire region of the complex system. Any variation of the probability density $\rho(\boldsymbol{r}, t)$ with t is accompanied by a probability density flow $\rho \boldsymbol{V}_D$ into or away from the respective field $\boldsymbol{r}$—Eq. (30).

The velocity field $\boldsymbol{V}_D(\boldsymbol{r}, t)$—Eq. (30), of the position probability changes with $\boldsymbol{r}$ and t like that of a non-differentiable hydrodynamic fluid in the force field of a potential $U(\boldsymbol{r}, t)$ and a non-differentiable potential $Q(\boldsymbol{r}, t)$. The non-differentiable fluid exhibits, however, an essential difference compared to an ordinary fluid: in a rotating motion $\boldsymbol{V}_D(\boldsymbol{r}, t)$ decreases (increases) with increasing (decreasing) from the center—(Eq. (9)).

Two types of complex system states are distinguished:

i. *Dynamic states*. For $\partial/\partial t = 0$ and $\boldsymbol{V}_D \neq 0$, that is, at the differentiable scale, Eq. (30) and Eq. (31) give:

$$\nabla\left(m_0 \frac{\boldsymbol{V}_D^2}{2} + Q + U\right) = 0 \tag{38}$$

$$\nabla \cdot (\rho \boldsymbol{V}_D) = 0 \tag{39}$$

specifically:

$$m_0 \frac{\boldsymbol{V}_D^2}{2} + Q + U = E \tag{40}$$

$$\rho \boldsymbol{V}_D = \nabla \times F \tag{41}$$

Consequently, the inertia, $m_0 \boldsymbol{V}_D \cdot \nabla \boldsymbol{V}_D$, the non-differentiable force, $-\nabla Q$ and the external force, $-\nabla U$, are balanced in every field point (Eq. (38)). The sum of the kinetic energy, $m_0 \frac{\boldsymbol{V}_D^2}{2}$, the non-differentiable potential, Q and the scalar potential, U, are invariant, *that is*, equal to the integration constant $E \neq E(\boldsymbol{r})$ (Eq. (40)). The parameter $E \equiv \langle E \rangle$ represents the total energy of the complex system dynamics. The flow density $\rho \boldsymbol{V}_D$ has no sources—Eq. (39), *that is*, its streamlines are closed—Eq. (41);

i. *Static states*. For $\partial/\partial t = 0$ and $\boldsymbol{V}_D = 0$, *that is,* at non-differentiable scale Eq. (30) and Eq. (31) give:

$$\nabla(Q + U) = 0 \tag{42}$$

i.e.,

$$Q + U = E \tag{43}$$

Consequently, the sum of the non-differentiable force, $-\nabla Q$, and the external one, $-\nabla U$, is null in every field point (Eq. (42)). Moreover, the sum of the non-differentiable potential, Q and the external one, U, is invariant, *that is*, is equal to the integration constant $E \neq E(\boldsymbol{r})$ (Eq. (31)). The parameter $E \equiv <E>$ represents the total energy of the static system. Eq. (7) is identically satisfied.

4. Self-multiplication and Taylor-type instabilities

In the following, using Eqs. (30) and (31) in a plane symmetry, we analyze, for example, the dynamics of the complex system, assuming that the fractal fluid is of isentropic type. Thus, we can choose the tensor (Eq. (32)) in the form $\sigma_{il} = \rho c^2 \delta_{il}$ [24], where ρ the density, c is the specific velocity, and δ_{il} is Kronecker's symbol. The existence of an external perturbation is determined by appropriate initial and boundary conditions (e.g., spatio-temporal Gaussian). In this context, we shall introduce the normalized coordinates

$$\omega t = \tau, kx = \xi, ky = \eta, \frac{V_{Dx}k}{\omega} = V_{\xi}, \frac{V_{Dy}k}{\omega} = V_{\eta}, \frac{\rho}{\rho_0} = N$$

where ω, k, and ρ_0 are critical parameters of the complex system (for details see [22, 23]).

Then, Eqs. (30) and (31) become:

$$\frac{\partial}{\partial \tau}(NV_{\xi}) + \frac{\partial}{\partial \xi}\left(NV_{\xi}^2\right) + \frac{\partial}{\partial \eta}(NV_{\xi}V_{\eta}) = -N^{-1}\frac{\partial N}{\partial \xi} \tag{44}$$

$$\frac{\partial}{\partial \tau}(NV_{\eta}) + \frac{\partial}{\partial \xi}(NV_{\xi}V_{\eta}) + \frac{\partial}{\partial \eta}\left(NV_{\eta}^2\right) = -N^{-1}\frac{\partial N}{\partial \eta} \tag{45}$$

$$\frac{\partial N}{\partial \tau} + \frac{\partial}{\partial \xi}(NV_{\xi}) + \frac{\partial}{\partial \eta}(NV_{\eta}) = 0$$

For numerical integration we shall impose the initial conditions:

$$V_{\xi}(0,\xi,\eta) = 0, V_{\eta}(0,\xi,\eta) = 0, N(0,\xi,\eta) = \frac{1}{5}, 1 \le \xi \le 2, 0 \le \eta \le 1 \tag{46}$$

as well as the boundary conditions:

$$\begin{aligned}
&V_{\xi}(\tau,1,\eta) = V_{\xi}(\tau,2,\eta) = 0, \quad V_{\eta}(\tau,1,\eta) = V_{\eta}(\tau,2,\eta) = 0 \\
&V_{\xi}(\tau,\xi,0) = V_{\xi}(\tau,\xi,1) = 0, \quad V_{\eta}(\tau,\xi,0) = V_{\eta}(\tau,\xi,1) = 0 \\
&N(\tau,1,\eta) = N(\tau,2,\eta) = \frac{1}{5} \\
&N(\tau,\xi,0) = \frac{1}{10}\, exp\left[-\left(\frac{\tau - \frac{1}{5}}{\frac{1}{5}}\right)^2\right] exp\left[-\left(\frac{\xi - \frac{3}{2}}{\frac{1}{5}}\right)^2\right] \\
&N(\tau,\xi,1) = \frac{1}{5}
\end{aligned} \tag{47}$$

By using the finite differences method [26], the system (Eq. (46)) with the initial conditions (Eq. (30)) and the boundary ones, Eq. (47) was solved numerically.

In **Figure 1a–c** [27], three-dimensional dependences of the normalized density N, normalized velocities V_{ξ} and V_{η}, on the normalized coordinates ξ and η are given for the normalized time τ = 0.54. Also, in **Figure 1d–f**, the contour curves of the

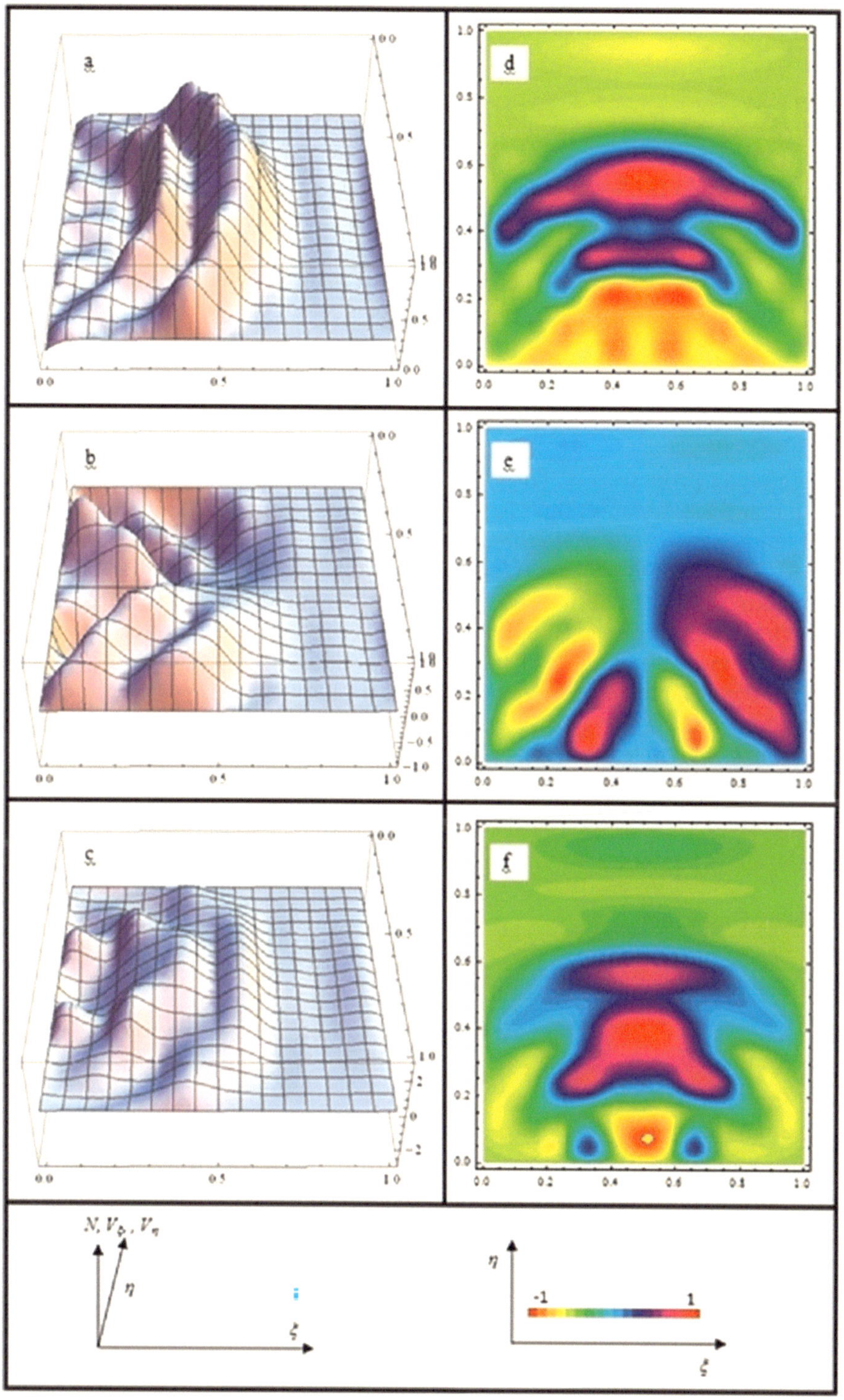

Figure 1.
a–f [27]. Three-dimensional dependences of the normalized density N*, normalized velocities* V_ξ *and* V_η*, on the normalized coordinates* ξ *and* η *for the normalized time* $\tau = 0.54$ *(a–c); two-dimensional contour of the normalized density* N*, normalized velocities,* V_ξ *and* V_η*, for the same normalized time (d–f).*

normalized density *N*, normalized velocities, V_ξ and V_η, are given for the same normalized time.

The following results: (i) the pattern generation; (ii) the symmetry of the normalized velocity, V_ξ, with respect to the symmetry axis of the spatio-temporal Gaussian; (iii) vertices at the pattern's periphery for the normalized velocity field, V_η (Taylor-type effect).

Since fractal/multifractal curves exhibit the property of self-similarity, then the dynamics of any complex system "mimic" holographic behavior (part reflects whole, whole reflects part).

Moreover, these holographic behaviors can be mapped to multilayered artificial neural networks of deep learning type [28].

5. Conclusions

This chapter proposes a unitary model for the dynamics of complex systems, positing that these systems might be equated to a multifractal mathematical object. Subsequently, when we explain these dynamics within the context of The Scale Relativity Theory, utilizing continuous and non-differentiable curves (fractal or multifractal), one might derive potential holographic behaviors. In this context, the Schrödinger and Madelung multifractal scenarios correlate through the generation of patterns, symmetries of the normalized velocity relative to the symmetry axis of the spatio-temporal Gaussian, and vertices at the periphery of the pattern for the normalized velocity field (Taylor-type effect). We offer a unified model for characterizing the dynamics of complex systems on both monofractal and multifractal manifolds. Our model is applicable to shape memory alloys, atmospheric conditions, biological fluids, etc.

Conflict of interest

The authors declare no conflict of interest.

Author details

Catalin Gabriel Dumitras[1]*, Vlad Ghizdovat[2], Cristina Maricela Rusu[3], Emanuel Nazaretian[1], Florin Nedeff[4], Valentin Nedeff[4], Iosif Ioja[4] and Maricel Agop[3]

1 Faculty of Machine Manufacturing and Industrial Management, Technical University Gheorghe Asachi, Iasi, Romania

2 Biophysics and Medical Physics Department, "Grigore T. Popa" University of Medicine and Pharmacy, Iasi, Romania

3 Physics Department, Technical University Gheorghe Asachi, Iasi, Romania

4 Faculty of Engineering, "Vasile Alecsandri" University, Bacau, Romania

*Address all correspondence to: catalin-gabriel.dumitras@academic.tuiasi.ro

References

[1] Bar-yam Y. Dynamics of Complex Systems, The Advanced Book Program. Reading, Boca Raton: Addison-Wesley, CRC Press; 1999. DOI: 10.1201/9780429034961

[2] Mitchel M. Complexity: A Guided Tour. Oxford: Oxford University Press; 2011. ISBN 9780199798100

[3] Flake GW. The Computational Beauty of Nature. Cambridge, MA: MIT Press; 2000. ISBN 9780262561273

[4] Buldyrev SV et al. Catastrophic cascade of failures in independent networks. Nature. 2010;**464**(7291): 984-985. DOI: 10.1038/nature08932

[5] Bennet CH, Yurek WH. How to define complexity in physics, and why, 2002. In: Gregersen NH, editor. From Complexity to Life. Oxford University Press; 2002. pp. 34-44. DOI: 10.1093/oso/9780195150704.003.0003

[6] Winfree AT. The Geometry of Biological Time. 2nd ed. New York: Springer; 2001. DOI: 10.1007/978-1-4757-3484-3

[7] Badii R, Politi A. Complexity: Hierarchical Structure and Scaline in Physics. Cambridge: Cambridge University Press; 1997. ISBN 9780521418904

[8] Mandelbrot BB. The Fractal Geometry of Nature. Updated and augm. ed. New York: W. H. Freeman; 1982. ISBN 0-7167-1186-9.

[9] Nottale L. Fractals and the quantum theory of spacetime. International Journal of Modern Physics A. 1989;**04**: 5047-5117. DOI: 10.1142/S0217751X89002156

[10] Nottale L. Scale Relativity and Fractal Space-Time; a New Approach to Unify Relativity and Quantum Mechanics. London: Imperial College Press; 2011. ISBN 978-1848166509

[11] Nottale L. Fractal Space-Time and Microphysics: Towards a Theory of Scale Relativity. Singapore: World Scientific; 1993. DOI: 10.1142/1579

[12] Bacaita ES et al. Nonlinearities in drug release process from polymeric microparticles: Long-time-scale behaviour. Journal of Applied Mathematics. 2012;**11**:1-26. DOI: 10.1155/2012/653720

[13] Munceleanu GV, Păun VP, Casian-Botez I, Agop M. The microscopic-macroscopic scale transformation through a chaos scenario in the fractal space-time theory. International Journal of Bifurcation and Chaos. 2011;**21**: 603-618. DOI: 10.1142/S021812741102888X

[14] Agop M et al. Informational non-differentiable entropy and uncertainty relations in complex systems. Entropy. 2014;**16**(11):6042-6058

[15] Agop M et al. On the information and uncertainty relation of canonical quantum systems with SL(2R) invariance. Chaos, Solitons & Fractals. 1996;7:659-668. DOI: 10.1016/0960-0779(95)00111-5

[16] Agop M et al. The uncertainty relation for an assembly of Planck-type oscillators. A possible GR-quantum mechanics connection. Chaos, Solitons & Fractals. 1997;**8**:809-821. DOI: 10.1016/S0960-0779(96)00101-4

[17] Agop M, Melnig Y. L'energie informationelle et les relations d'incertitude pour les systemes

canoniques SL(2R) invariants. Entropie. 1995;**188**(189):119-123

[18] Gurlui S et al. Experimental and theoretical investigations of transitory phenomena in high-fluence laser ablation plasma. Physical Review E. 2008;**78**:026405. DOI: 10.1103/PhyRevE.78.026405

[19] Agop M, Paun V, Harabagiu A. El Naschie's $\varepsilon(\infty)$ theory and effects of nanoparticle clustering on the heat transport in nanofluids. Chaos, Solitons & Fractals. 2008;**37**:1269-1278. DOI: 10.1016/j.chaos.2008.01.006

[20] Phillips AC. Introduction to Quantum Mechanics. New York: Wiley; 2013. 288 pp. ISBN 78-1-118-72325-8

[21] Bohm D. Quantum Theory. New York: Dover Publ; 1989. ISBN 978-0486659695

[22] Michel OD, Thomas BG. Mathematical Modeling for Complex Fluids and Flows. Springer; 2012. 264 pp. ISBN 978-9814273251. Available from https://doi.org/10.1007/978-3-642-25295-2

[23] Thomas YH. Multi-Scale Phenomena in Complex Fluids: Modeling, Analysis and Numerical Simulations. World Scientific Publishing Company; 2009. ISBN 978-9814273251. DOI: 10.1142/7291

[24] Landau LD, Lifshitz EM. Fluid Mechanics. 2nd ed. Pergamon Press; 1987. ISBN: 978-0080339337. Available from: https://doi.org/10.1016/C2013-0-03799-1

[25] Batchelor GK. An Introduction to Fluid Dynamics. Cambridge: Cambridge University Press; 2000. 615 pp. DOI: 10.1017/CBO9780511800955

[26] Zinkiewikz OC, Taylor RL, Zhu JZ. The Finite Element Method: Its Basis and Fundamentals. 6th ed. Elsevier Butterworth-Heinemann; 2005. 714 pp. ISBN 978-1856176330 Available from https://doi.org/10.1016/C2009-0-24909-9

[27] Nejneru C et al. Dynamics control of the complex systems via nondifferentiability. Journal of Applied Mathematics. 2013;**1**:137056

[28] Shimobaba T, Blinder D, Birnbaum T, Hoshi I, Shiomi H, Schelkens P, et al. Deep-learning computational holography: A review. Frontiers in Photonics. 2022;**3**:854391

Chapter 5

Agent-Based Models

Boris Atenas, Edward Larroza and Verónica Bahoz

Abstract

This chapter explores agent-based models (ABMs) as computational tools for simulating complex systems through autonomous agents and their interactions. We discuss the foundational principles of ABMs and their connection to dynamic systems theory. Applications across socio-physics, such as opinion dynamics, civil unrest, fake news, urban transportation policies, and education, illustrate how micro-level interactions lead to macro-level societal patterns, including Pareto's laws. In biology, ABMs have proven valuable in modeling medical treatments, such as magnetic fluid hyperthermia, as well as in simulating the spread of diseases like COVID-19. Additionally, dynamical systems such as Lotka-Volterra, when combined with ABMs, can offer complementary insights into complex interactions. Finally, in the economics section, we examine market behaviors, wealth distribution, and strategic interactions, such as the Minority Game. We expect that with these examples, the reader will gain a deeper understanding of how ABMs provide a powerful framework for capturing the emergent behaviors of complex systems across diverse fields.

Keywords: socio-physics, complex systems, economy, politics, social networks, education, biological systems, Ising model

1. Introduction

An agent-based model (ABM) is a computational simulation technique that represents complex systems through the interaction of autonomous agents. Each agent is an independent entity with specific characteristics, behaviors, and rules for interacting with other agents and the environment. In these interactions, individuals share common characteristics, such as opinions, emotions, decision-making, and more. When these opinions or behaviors propagate to neighboring individuals, a dynamic emerges that is often characterized by complex system behavior. In such systems, external factors like policies or rule changes act as forces that alter the system's dynamics. Historically, this type of interaction has been observed in both biological systems—such as bacterial colonies influenced by external factors like electric or magnetic fields and chemical substances, and in the collective behaviors of animal species, such as schools of fish, flocks of birds, swarms, and herds—and social systems are characterized by complex interactions that give rise to collective human behaviors, such as the organization and development of cities, tribes, clans, and various other social groups. These interactions can be observed across different scales of human society, from small familial units to larger, more structured communities like gangs or entire cities, illustrating the fundamental processes behind the formation and evolution of social structures. In contrast with biological systems, as humans, we have developed diverse areas of knowledge and activities, which means that our interactions extend into vastly different fields, including education, sports, economy, politics, and social networks. Thus, it seems natural to model the behavior of these varied collectives through simulations that capture individual interactions while accounting for external forces that influence system dynamics.

In socio-physics, most agent-based models take inspiration from the Ising model [1] as a foundational framework for studying complex systems. Galam [2] was one of the pioneers in applying such models to social systems, where individual interactions are represented similarly to the alignment of spins in magnetic materials, offering new insights into collective behaviors. The Ising model is particularly significant as it was originally developed to explain ferromagnetism. Ferromagnetic materials have intrigued humanity with their "magical" properties since ancient times, but it was not until scientific exploration revealed the interactions between the underlying elements of matter that this phenomenon began to be understood. Later, the discovery of the concept of spin brought the Ising model to the forefront of ferromagnetism research. With further advancements by scientists, such as Heisenberg, Bloch, Frenkel, and Landau, ferromagnetism was eventually explained in more detail through first principles, solidifying the Ising model's relevance in this field.

The Ising model assumes a system of N particles with non-zero intrinsic magnetic moments arranged on a lattice. In the one-dimensional case, this forms a chain, while in two dimensions, it forms a square grid. The magnetic moments, called "spins," align along the z-axis and take values +1 or − 1. When spins are randomly distributed, the system exhibits paramagnetic behavior with no net magnetization. If the spins show some order, ferromagnetic behavior arises. The Hamiltonian for the Ising model is given by

$$\mathcal{H} = -J\sum_{i,j} S_i S_j - h\sum_{i} S_i, \tag{1}$$

where J represents the interaction between spins, and h is an external magnetic field. Years later, a generalization of Ising model, the "Potts model" [3], was introduced, allowing spins to adopt more states beyond just "up" and "down." This model has gained even wider usage, as it provides individuals with more options when making choices. Given their adaptability, the Ising and Potts models have found applications in various fields.

Dynamic systems theory plays a fundamental role in understanding how micro-level interactions evolve into macro-level patterns and behaviors by studying how a system's state changes over time. This is typically achieved through differential equations that describe continuous changes in variables, often represented by vectors that capture both the direction and magnitude of these changes in a multi-dimensional space. By analyzing these equations, it is possible to predict long-term behavior, trace system trajectories, and identify points of stability or instability influenced by both internal interactions and external forces." For example, Hamiltonian systems are a class of dynamic systems particularly useful in economic modeling. They describe the evolution of variables over time under the influence of external factors using Hamilton's equations:

$$\dot{q} = \frac{\partial H}{\partial p}, \quad \dot{p} = -\frac{\partial H}{\partial q} \tag{2}$$

where $H(p, q, t)$ is the Hamiltonian function, q represents the state variables, and p is the conjugate momenta. These equations allow economists to model the optimal allocation of resources over time, capturing trade-offs between immediate consumption and future investment. By integrating Hamiltonian dynamics with agent-based models, it is possible to explore how individual agent behaviors aggregate to influence macroeconomic growth and development.

Similarly, the Lotka-Volterra equations are fundamental in modeling biological systems, especially in ecology. They describe the dynamics of predator-prey interactions and competition between species through a set of differential equations:

$$\frac{dN}{dt} = N(a - bP), \frac{dP}{dt} = P(cN - d) \tag{3}$$

where N and P represent the populations of prey and predators, respectively, and $a, b, c,$ and d are parameters that represent interaction rates. These equations model how populations evolve over time due to births, deaths, and interactions and are studied using vector representations in phase space.

By connecting agent-based models with dynamic systems, we can better understand how micro-level interactions between individual agents lead to macro-level behaviors within the system. While ABMs provide detailed simulations of individual actions, dynamic systems offer the mathematical tools to study the overall evolution and long-term dynamics of these systems.

Aware that agent-based models have significantly broader applications in various other fields, and while both economic and biological systems can be studied within the broader framework of socio-physics, in this chapter, we distinguish them as separate categories to highlight their unique applications and methodologies: socio-physical, economic, and biological systems.

2. Socio-physics

Socio-physics is an interdisciplinary field that applies principles from physics to study and understand social phenomena. By modeling individuals as interacting particles or agents, socio-physics seeks to uncover the underlying mechanisms that drive collective behaviors in societies. In this section, we explore a variety of agent-based models that simulate complex social dynamics, including opinion formation, civil unrest, education, and urban transportation patterns.

2.1 Opinion dynamics

Within socio-physics, several models have been developed to study the complex dynamics of social behavior and opinion formation. One such prominent model is the Sznajd model, which focuses on opinion dynamics and has been extensively applied in fields such as sociology, statistical physics, and network theory to study how opinions evolve within a group of interacting individuals [4–8]. It provides a simplified yet powerful framework for understanding how consensus is formed or how communities become polarized through local interactions between individuals. The model relies on the principle of social validation, where people tend to align their opinions with those of their neighbors, resulting in large-scale patterns of agreement or disagreement over time.

In this model, the agents are represented as "spins" that can either be in the "up" or "down" state, interact locally with their neighbors. These spins symbolize the opinions of individuals, and the interactions between spins reflect the social influence that each individual exerts on others within their local vicinity. The dynamics of the model can be controlled by different rules of interaction, which govern how an agent changes its opinion based on the opinions of its neighbors.

The model has been used to explore how small local interactions can lead to large-scale social phenomena, such as the formation of public opinion, political polarization, or even the sudden shifts in collective behavior known as phase transitions in physical systems. Through simulations, how different initial conditions and rules of interaction affect the speed and nature of opinion convergence or divergence within a population can be investigated. Therefore, the Sznajd model offers a rich theoretical framework for studying complex social behaviors. The presented figure highlights how varying the initial proportion of opinions and modifying the rules of interaction can lead to significantly different patterns of consensus or fragmentation within a community. Such insights are valuable in fields ranging from political science to marketing, where understanding how opinions spread and change is of critical importance.

2.2 Social networks

Social networks have become pivotal in shaping public discourse, but they also facilitate the rapid spread of misinformation, commonly referred to as "fake news." Understanding how fake news propagates through these networks is essential for developing effective countermeasures. Although our coverage of social networks is brief, it is important to highlight how agent-based models contribute to this field.

Burbach et al. [9] utilized an agent-based model (ABM) to simulate the dynamics of message dissemination in online networks. Their model incorporated personality traits such as extraversion, narcissism, and emotional self-efficacy to determine how these factors influence an individual's likelihood of sharing fake news. The simulations revealed that fake news spreads more widely and quickly than regular news, especially in highly connected networks. In their model, up to 39% of agents acted as spreaders in a network of 1000 agents, underscoring the significant role of network structure and user behavior in amplifying misinformation.

Complementing these findings, Zhao et al. [10] conducted an empirical study analyzing the real-world spread of fake news on social media platforms like Weibo and Twitter. Their research demonstrated that fake news propagates faster and reaches a broader audience than true information, primarily due to network structural characteristics such as higher connectivity and the influence of key nodes. Although this study did not employ an ABM, it confirmed insights from the simulations—namely, that fake news benefits from network configurations that accelerate its dissemination.

Together, these studies offer complementary perspectives: the agent-based model by Burbach et al. provides micro-level insights into how individual traits influence the spread of fake news, while the empirical study by Zhao et al. emphasizes the macro-level impact of network structures. Both conclude that fake news not only spreads faster but also reaches a wider audience than real news, highlighting the significant roles of individual behavior and network topology.

This underscores the utility of ABMs in simulating complex social phenomena and validating them against real-world data. By modeling individual behaviors and interactions within a network, ABMs offer valuable insights into the mechanisms behind the rapid spread of misinformation, informing strategies to mitigate its impact.

2.3 Civil violence

Civil violence arises from complex interactions between social dissatisfaction and authority responses, often driven by the tension between government legitimacy and public unrest. To better understand these dynamics, the Epstein model [11] simulates how protests escalate or are subdued based on key factors such as police presence and the perceived risk of punishment. Epstein demonstrated complex phenomena such as the emergence of deceptive behaviors in citizens, who pretend to be calm when police are present but rebel when authorities are absent. This model is essential for understanding the mechanisms that lead to civil unrest and how strategic interventions, like deploying police forces, can either prevent or exacerbate conflict escalation. Since the introduction of Epstein's model, several authors have expanded its scope. Among them, Lemos et al. [12] analyzed the decision rules governing the agents' behaviors, while Ormazábal et al. [13] explored the influence of money distribution on civil unrest. Recently, Ormazábal et al. [14] applied a statistical physics approach to extend Epstein's model by introducing a Potts-type energy function:

$$E[\alpha] = -\frac{1}{2N}\sum_{i,j} J_{ij}\,\delta(\alpha_i, \alpha_j), \tag{4}$$

where α_i and α_j represent the states of agents i and j, J_{ij} is the interaction strength between the agents, and $\delta(\alpha_i, \alpha_j)$ is the Kronecker delta, which equals 1 when the agents are in the same state (e.g., both passive or both active) and 0 otherwise. This function allows us to quantify the system's stability: when the energy is low, the system tends to an ordered state, such as when most agents are passive.

In this model, the parameter T acts as a threshold analogous to temperature in magnetism models like the Ising model. At low values of T, agents are less likely to switch states, leading to an ordered and stable system where consensus (i.e., a majority of passive agents) prevails. As T increases, agents are more likely to become active, introducing disorder into the system, much like how higher temperatures in the Ising model disorder spins.

Ormazábal et al. further showed that the **Principle of Minimal Aggravation** governs the model's dynamics: agents shift from passive to active based on their dissatisfaction with government legitimacy and the perceived risk of arrest. They developed a phase diagram that illustrates transitions between ordered and disordered phases, offering a deeper understanding of how factors such as legitimacy and police presence influence protest dynamics.

Figure 1 illustrates how legitimacy and activation thresholds affect the evolution of protests. The left panel is critical as it clearly depicts the moments when civil unrest may escalate or be controlled, depending on the balance of these key variables. It shows the phase diagram of the system without police, with transitions between the AP phase (dominated by active agents), the PA phase (dominated by passive agents), and the consensus phase, where all agents are passive. These transitions are governed by the legitimacy of the government and the activation threshold T. As legitimacy increases and threshold T decreases, the system tends to reach a more ordered state

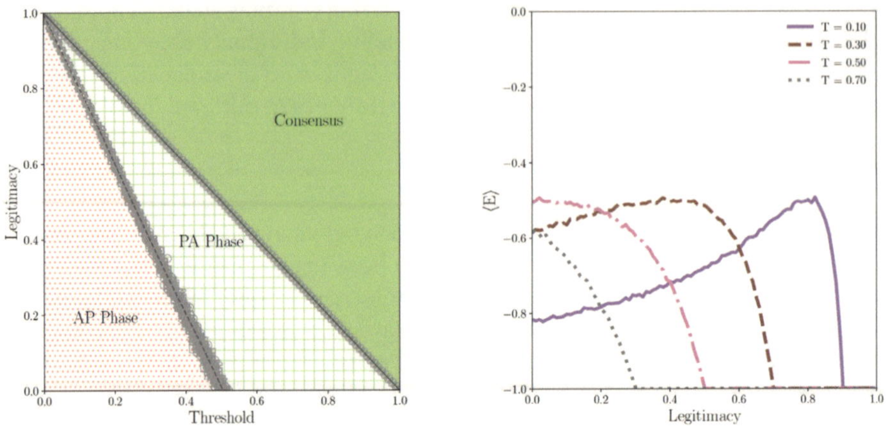

Figure 1.
The left panel shows the phase diagram of the system without police, illustrating the transitions between the AP phase (active agents), PA phase (passive agents), and the consensus phase, where all agents are passive, as a function of legitimacy and threshold (T). The right panel presents the global average energy E variations for different values of threshold T, demonstrating the system's stability transitions based on legitimacy. Figures cortesy of Ormazabal.

(consensus). Conversely, lower legitimacy or higher threshold values lead to disordered states, where protests are more likely to occur.

The right panel complements this by presenting the global average energy E variations for different values of threshold T. This graph illustrates how the system's energy, which serves as an indicator of disorder, changes as a function of legitimacy. For lower threshold values, the system stabilizes at lower energy levels, corresponding to more ordered states. As T increases, the energy rises, signaling more disordered configurations, with active agents becoming predominant. The two curves together highlight the interplay between legitimacy, threshold, and system stability, revealing how these factors collectively determine whether a system remains stable or descends into civil unrest.

2.4 Urban transportation polices

The literature on transportation systems offers valuable insights through the use of ABMs to evaluate the potential impacts of public policies. Maggi and Vallino [15] applied ABMs to explore urban mobility and assess the effects of transportation policies, highlighting the capability of these models to capture complex interactions among different actors in the system. However, they also noted that real-world applications of these models remain limited.

Building upon this foundation, Zhuge et al. [16] employed an agent-based spatial integrated framework to examine the future electric vehicle (EV) market in Beijing. Their study assessed environmental benefits such as emission reductions, analyzed neighborhood effects on EV adoption rates, and evaluated the potential strain on the electric grid. They found that by 2020, EVs would account for only 4% of total electricity demand, underscoring the need for expanded charging infrastructure to support further growth.

Expanding the scope to urban planning, Wise et al. [17] explored the integration of transportation systems into agent-based urban models, highlighting the complexities of urban mobility. Their research emphasized the importance of developing more sophisticated models capable of accounting for critical factors such as accessibility, transportation costs, and network effects on city planning and infrastructure.

Similarly, Stankov et al. [18] applied ABMs to study commuter behavior and transportation patterns in Bogotá, taking into account factors such as spatial accessibility, personal safety, and social influences. Their study examined how transportation fare subsidies promote the use of public transport, particularly among low-income groups. The study also examined the impact of these policies on physical activity. Although the policies did not reduce congestion, they decreased car usage among middle- and high-income groups. The results are illustrated in **Figure 2**. Left chart; Left bar: Represents actual data from the 2019 Bogotá household survey. Right bars: Represent the simulated results from the model. Right chart: Shows the absolute percentage change in the use of each mode of transportation before and after the implementation of the policies (simulation only). Panel A: Shows the change in transportation use following a 30% fare subsidy. Panel B: Shows the change in transportation use with free public transport. Panel C: Shows the change in transportation use with a congestion charge of 2000 COP. Panel D: Shows the change when combining a 30% subsidy with a 2000 COP congestion charge. Panel E: Shows the change with a 20,000 COP congestion charge. Panel F: Shows the change when combining free public transport with a 2000 COP congestion charge. Panel G: Shows the change when combining a 30% subsidy with a 20,000 COP congestion charge. Panel H: Shows the change when combining free public transport with a 20,000 COP congestion charge.

These results highlight not only the immediate effects of transportation policies but also suggest potential long-term shifts in commuter behavior, especially in terms of how subsidies and congestion taxes can influence mode share across different socioeconomic groups. While fare subsidies significantly increase public transportation use among low-SES groups, the combination with congestion taxes leads to a

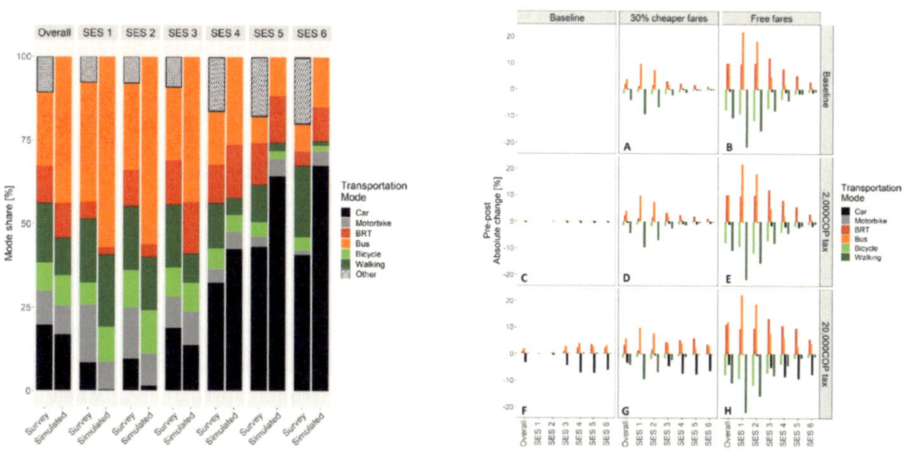

Figure 2.
(Left) Comparison of actual transportation mode share from the 2019 Bogotá Household Travel Survey with model-simulated results across socioeconomic groups (SES). (Right) Absolute percentage change in transportation mode share following the implementation of various fare subsidies and congestion charges. Figures cortesy of Stankov.

broader reduction in car usage, particularly in mid-to-high-SES groups. This dual approach may foster a more sustainable transportation system, reducing both congestion and social inequities in access to transportation.

2.5 Education

Educational environments are complex adaptive systems with similar rules to those described in other systems. Here, teachers and students interact to achieve their goals, acquiring useful knowledge to perform specific tasks in their respective fields of study, thereby forming professionals and experts in undergraduate, master's, and doctoral programs. Agent-based models have been used to explore topics such as university organizational management and strategy, inter-institutional collaboration, academic processes like recruitment and peer review, student enrollment and specialization choices, academic performance, measurement of student workload, curriculum redesign, and more [19–24].

One of the first works in the literature on teaching-learning processes is that of Bordogna et al. [20], where a multidisciplinary approach is applied to describe social interactions between individuals in the educational system and behavioral processes. Using Monte Carlo simulations of a generalized Ising model and defining a variable called "cognitive impact" (the teacher's ability to persuade and capture the student's attention), the interactions between Teacher-Student, interactions among students, and the interaction of students with study materials are modeled. In this model, the knowledge of the students is modeled as a dynamic variable that changes based on the total cognitive impact from all sources of knowledge (teacher, peers, and study materials). The Metropolis algorithm is used to model the probability that a student improves their knowledge over time. The results of this work showed different time regimes where high-performing and low-performing students benefit from different group configurations (homogeneous and heterogeneous). In later work, Ormazabal et al. [22] developed an agent-based model that considers dynamic interactions, meaning the interactions are no longer fixed but can change. This allows for observing how specific changes in teaching strategies affect learning. It includes real-time interaction and students' responses to different pedagogical approaches, which can vary and adapt during the learning process. A significant aspect of this study is that it models how student attention influences the effectiveness of educational interactions. It considers the possibility that a student, in the middle of the process, becomes more motivated, increasing their attention to the course, which translates into improved performance. This is important because attention can vary depending on many factors, including the teaching method, the level of interaction with the teacher, and group dynamics, among others.

Figure 3 shows a clear phase transition-like behavior in both homogeneous and heterogeneous classes, driven by the noise parameter β . As β increases, which represents the level of uncertainty or distractions in the learning environment, the amount of knowledge acquired by the students (σ) decreases. The critical point, β_c , marks a threshold beyond which learning efficiency drops significantly, mimicking a transition from an ordered (high-knowledge acquisition) to a disordered state (low-knowledge acquisition). In the homogeneous class (left panel), the transition is more abrupt, with students showing a steeper decline in performance as noise increases. The heterogeneous class (right panel) exhibits a similar trend, but the distribution of σ across different cases suggests that the impact of noise is more varied due to the

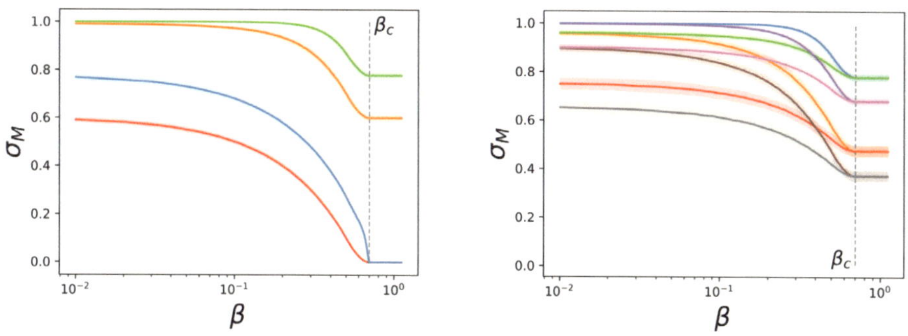

Figure 3.
Homogeneous Class vs. Heterogeneous Class. On the left, for a homogeneous class, and on the right, for a heterogeneous class. The amount of noise (β) *and the amount of knowledge acquired* (σ) *are displayed. In Case I, students attended only lectures; in Case II, they attended lectures and participated in group work; in Case III, they attended lectures and engaged in individual study; and finally, in Case IV, they attended lectures, participated in group work, and engaged in individual study. The results show that the more learning activities students engage in, the greater their knowledge acquisition, while higher noise levels reduce their learning outcomes.* **Table 1** *shows the cases studied for the heterogeneous class. Figures were provided by Ormazabal.*

diversity in learning interactions. This highlights how increasing learning activities can buffer against noise to some extent but also emphasizes the importance of managing distractions to prevent a collapse in the learning process, especially when the noise exceeds the critical threshold β_c.

Recently, Velázquez and his collaborators [23–25] developed an ABM using empirical data, providing a detailed look at the teaching-learning process in the context of online education, with a focus on student workload through the analysis of activity logs on digital platforms.

In these studies, it was observed that students' daily dedication to learning activities is unequal and follows a Pareto or Zipf laws [23] (see left side of **Figure 4**), where an assessment serves as an incentive that triggers an increase in the number of hours students dedicate to their studies, which are related to other patterns of human behavior. Additionally, performance indices were introduced, such as student progression, accumulation rate, over-duration, and learning rate, among others, which are used to track and assess the performance of both students and teachers.

The cumulative distribution of final grades in various courses with a significant number of students is described in **Figure 4** (right side), where standardized evaluations, such as written tests, are applied. It was observed that these distributions exhibit a regular pattern, with a uniform distribution in the failing zone and an almost Gaussian distribution in the passing zone. Using that empirical data, Monte Carlo simulations were carried out through an agent-based model that incorporates dropout rules for both students and the educational institution [24], which results successfully replicated the empirical grade distribution of students, capturing both the Gaussian behavior for passing grades and uniform distribution for failing ones, offering new insights into optimizing student workload [24].

Among the results obtained, a mechanism is established to estimate the time required for students to complete a specific course based on their academic performance. Additionally, these models are capable of reproducing performance indicators of real processes, such as the pass rate and average grades, providing a powerful tool for optimizing both student workload and institutional planning.

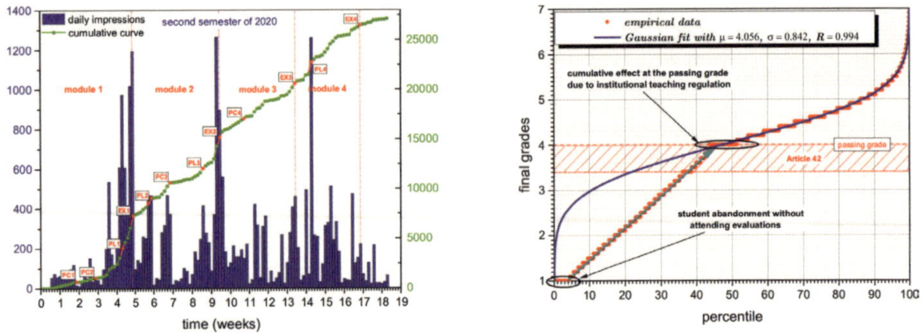

Figure 4.
(Left) Daily YouTube impressions. (Right) Empirical data of cumulative distribution function of final grades in the course "Introduction to Physics" of a Chilean University.

Cases	Class attendance	Group work attendance	Individual study
OOO	Obligatory	Obligatory	Obligatory
OOR	Obligatory	Obligatory	Random
ROO	Random	Obligatory	Obligatory
ROR	Random	Obligatory	Random
ORO	Obligatory	Random	Obligatory
ORR	Obligatory	Random	Random
RRO	Random	Random	Obligatory
RRR	Random	Random	Random

[a]*Obligatory means full attendance to the activity.*
[b]*Random indicates attendance depends on the student's choice.*

Table 1.
Different cases of class attendance, group work attendance, and individual study.

3. Biological systems

In biological systems, the ABM provides a flexible and powerful framework to simulate complex interactions at the individual level. By representing cells, organisms, or particles as autonomous agents governed by specific rules, ABMs allow capture emergent phenomena that traditional models often fail to address. This approach has been instrumental in advancing our understanding of processes such as cellular behavior, disease propagation, and therapeutic interventions.

One of the early recognition of the potential of ABMs in biology was highlighted by Pleyer and Fleck [26]. They emphasized that ABMs offer a bottom-up approach to modeling multicellular systems, where individual cells are treated as agents interacting within dynamic environments. This method captures emergent properties such as pattern formation, cellular migration, and self-organization, which are difficult to model using continuous approaches like partial differential equations (PDEs). By simulating cells with distinct behaviors and local interactions, ABMs have shed light on the complexities of biological pattern formation and tumor dynamics, contributing significantly to the field of systems biology.

Other interesting work was performed by Wei et al. [27], who developed a multi-agent system (MAS) to simulate the decentralized processes of memory formation and retrieval in biological neural networks. Traditional models often assume centralized control mechanisms, but Wei et al.'s MAS model uses autonomous agents that interact locally to form memory paths. By integrating graph theory and physical laws such as Ohm's and Kirchhoff's laws, the system emulates neuron-like behavior, particularly the reinforcement of memory traces through repeated stimulation. Their results demonstrated that memory could be successfully encoded, stored, and retrieved through these local interactions without a global controller. This work offers valuable insights into the neurobiological mechanisms underlying memory processes.

The advent of the COVID-19 pandemic spurred further applications of ABMs in epidemiology. Recently, Schinko et al. [28] developed an ABM to simulate the airborne spread of SARS-CoV-2 in indoor environments. Recognizing that virus transmission is influenced by both individual behaviors and environmental factors, they integrated their ABM with Computational Fluid Dynamics (CFD) to account for aerosol dispersion. In this model, agents represent individuals moving through a space and performing actions like coughing or speaking, which release aerosols into the environment. The movement and behavior of these agents serve as dynamic inputs that influence the flow of virus-laden particles, creating a complex, evolving system. By combining ABMs with CFD, the study captured the interactions between individual behaviors and environmental conditions, providing valuable insights into how movement patterns, ventilation, and social distancing affect virus transmission in enclosed settings. This highlights the role of ABMs in modeling dynamic systems where agent interactions and environmental factors evolve over time, offering a powerful framework for assessing risks and designing effective intervention strategies.

Another significant contribution to medical applications came from Fernandes and Miranda [29], who developed an agent-based computational model for magnetic fluid hyperthermia (MFH). MFH is a cancer treatment technique where magnetic nanoparticles (MNPs) are used to locally heat tumor tissues. In their model, MNPs are represented as autonomous agents within a discretized environment divided into patches, each with its own temperature. When activated by an alternating magnetic field, these nanoparticles generate heat through Brownian relaxation (physical movement of the MNPs) and Néel relaxation (rotation of the magnetic moments within the particles). The temperature evolution in each patch is calculated based on the energy provided by the MNPs and heat transfer between neighboring patches following Fourier's law, using a von Neumann neighborhood scheme.

The model progresses in discrete steps, simulating the random movement of MNPs and their impact on the spatial distribution of heat. When comparing simulations with experimental data from Fortin et al. [30] on glioblastoma multiforme treatments, the model adequately reproduced the non-linear behavior of temperature increase relative to MNP concentration and the thermal stabilization of the system. The results indicated that internal parameters, such as the coefficient α (related to thermal conductivity and specific heat), have a greater influence on temperature stabilization and heating speed than external parameters like the intensity of the magnetic field. Additionally, local heat exchange interactions between patches and MNPs are key to achieving homogeneous temperature within tumor tissue. This model provides a valuable tool for understanding the factors affecting thermal stability in MFH treatments and could contribute to optimizing treatment parameters in personalized clinical applications.

Another key tool in the study of biological systems is the Lotka-Volterra model [31], which has been widely used to explore predator-prey interactions, offering a simple yet powerful framework for understanding the dynamics of competing populations. This model, which captures the oscillations in predator and prey populations due to their interactions, is essential in ecological modeling. While the classical Lotka-Volterra model provides valuable insights into these dynamics, it can be complemented by agent-based simulations (ABMs), which allow for the inclusion of additional factors that are difficult to capture in traditional dynamic system models—such as individual-level variability, spatial distribution, or adaptive behavior. In this way, ABMs can extend the scope of ecological studies by simulating more complex scenarios, like saturation effects or chaotic dynamics, providing a richer understanding of the processes that drive ecological systems. In fact, the Lotka-Volterra framework has also been extended to other fields, such as economics, where it helps explain wealth and income distributions. The generalized Lotka-Volterra (GLV) models, combined with agent-based simulations (ABMs), have been applied to study the dynamics of capital accumulation and income inequality, demonstrating that fundamental economic drivers, such as capital concentration, can produce distributions observed in real-world economies. This highlights the flexibility of the Lotka-Volterra framework and the value of ABMs in capturing emergent behaviors in complex systems, whether in biological or economic contexts [32]. Another example of combining both techniques is seen in a variant of the Lotka-Volterra model that introduces three-agent interactions, inspired by cyclic competition like the Rock-Paper-Scissors game. This approach explores how cooperative predation and stochastic dynamics influence species coexistence and the formation of spatial patterns, offering richer insights beyond the traditional two-species interactions [33].

In summary, Agent-Based Models have profoundly influenced the study of biological systems by enabling the simulation of complex, dynamic interactions that are difficult to capture with traditional modeling approaches. From cellular behavior and neural networks to epidemiology and cancer treatment, ABMs have provided insights that enhance our understanding of biological processes and support the development of effective interventions. By focusing on individual agents and their local interactions, these models reveal how micro-level behaviors can lead to emergent macro-level phenomena, reinforcing the importance of ABMs in advancing biological research.

4. Economics

In the study of economics, the ABMs have become essential tools for simulating a wide range of market behaviors, from the intricacies of order books and wealth distribution to strategic interactions exemplified by the Minority Game [34, 35]. These models enable researchers to explore complex economic phenomena by capturing the interactions of individual agents, whose collective behavior often leads to emergent properties not readily apparent from traditional analytical models.

One of the early breakthroughs in modeling market microstructure using ABMs was the work of Lux and Marchesi [36], who introduced a framework in which fundamentalists, chartists, and noise traders interact within a market environment. In their model, price variations emerge directly from excess demand: at each time step, agents submit their orders, and the resulting price is recalculated based on market imbalance. This approach allowed for a dynamic view of market prices, capturing key features such as volatility clustering and fat-tailed return distributions. However, it lacked the concept of a persistent order book for tracking trades, which limited its ability to simulate the depth and liquidity of real markets.

Proceeding from this foundation, Chiarella and Iori [37] expanded the model by introducing a persistent order book where agents place limit orders based on different trader profiles. This addition allowed for a more realistic simulation of liquidity and price formation, as limit orders are a fundamental aspect of modern financial markets. By incorporating a continuous record of unexecuted orders, their model could capture the dynamics of order accumulation and depletion, providing deeper insights into market microstructure.

Further advancing the empirical realism of ABMs in finance, Mike and Farmer [38] developed one of the most empirically grounded models of order books to date. Their model incorporated detailed processes of order placement and cancelation, fitted to real market data. By closely aligning the model's parameters with observed trading behavior, they bridged the gap between theoretical models and the practical, observed dynamics of financial markets, enhancing the predictive power and applicability of ABMs in economic research.

Parallel to these developments in modeling market microstructure, ABMs have also been instrumental in exploring wealth distribution within economies. Dragulescu and Yakovenko [39] applied concepts from statistical mechanics to study money distribution, demonstrating that it follows a Boltzmann-Gibbs exponential distribution in closed economic systems. This approach treated money exchange between agents analogously to energy exchange between particles, providing a novel perspective on economic interactions.

Expanding on this idea, Chakraborti and Chakrabarti [40] introduced a model examining how agents' saving propensities affect wealth distribution. They showed that incorporating a uniform saving propensity among agents leads to a gamma-like distribution of wealth, highlighting the role of individual behavior in macroeconomic outcomes.

A significant leap was made by Mohanty [41], who presented an agent-based model that successfully reproduced the Pareto law for wealth distribution. In Mohanty's model, agents exchange wealth in a manner similar to particles exchanging energy in an ideal gas, but with the crucial addition of individual saving propensities. This feature was key in generating a Pareto-like distribution for the wealthiest agents, while the majority of the population followed an exponential or Gibbs distribution. This dual behavior mirrored real-world observations, where wealth distributions often exhibit a Pareto tail alongside a bulk exponential distribution.

Building on these foundations, more recent studies such as that of Aydiner et al. [42] have demonstrated the robustness of Pareto distributions in agent-based models, even when accounting for diverse conditions like initial wealth distributions and debt. The flexibility of these models illustrates how wealth inequality can persist across various economic environments, emphasizing the importance of microscopic interactions in shaping macroeconomic patterns.

In addition to wealth distribution and market microstructure, ABMs have been applied to study strategic interactions and market manipulation. For example, Fratric et al. [43] used an ABM to investigate market manipulation in the Bitcoin market during the 2017–2018 boom. Their study revealed how a fraudulent agent could significantly influence market prices, demonstrating the crucial role of liquidity and agent behavior in amplifying or mitigating manipulative actions. This work highlights the potential of ABMs to uncover vulnerabilities in financial systems and inform regulatory policies.

Transitioning from wealth dynamics and market manipulation to strategic decision-making, the Minority Game stands out as a powerful tool for simulating competitive environments where agents with limited information vie to be in the minority. Originating from the El Farol Bar Problem introduced by Arthur [44], the Minority Game models situations where agents benefit from being in the minority group. In the El Farol Bar Problem, individuals decide whether to attend a bar that becomes overcrowded if too many people go, making it more enjoyable for those who stay home and vice versa.

Challet and Zhang [45] formalized this concept into the Minority Game, where agents must repeatedly choose between two actions—such as buying or selling assets—without knowing what the majority will do. The goal is to be in the minority, as those agents gain a reward while the majority loses. The game introduces the idea of bounded rationality: agents base their decisions on historical data and adaptive strategies rather than perfect information.

As agents participate in each round of the game, they adjust their strategies over time, discarding those that consistently lead to failure. This dynamic behavior mimics real-world financial markets, where traders constantly refine their strategies based on past performance. Challet and Zhang [46] introduced an evolutionary version of the game, where underperforming agents are replaced by "clones" of more successful agents. This Darwinian process allows the system to "learn" and improve overall performance, paralleling the evolutionary dynamics observed in financial markets.

Further developments in the Minority Game have explored the dynamics of cooperation and network interactions. Zhang et al. [47] investigated how allowing agents to dynamically rewire their connections based on cooperation success can enhance overall system efficiency. Their model demonstrated that adaptive network structures significantly improve cooperation, especially in small-world and random networks, where decision-making becomes more stable and efficient.

Building on the concepts of agent competition and strategy diversity, Xu and Zhong [48] examined how heterogeneous investment strategies and reference points contribute to market stability. Their findings indicate that the coexistence of diverse investor types prevents extreme price fluctuations, providing a valuable extension to traditional Minority Game models and reinforcing the importance of diversity in economic systems.

A recent real-world example that mirrors the dynamics explored in the Minority Game is the 2021 GameStop stock event. Here, retail investors coordinated via online platforms to drive up the stock price of GameStop, creating a massive short squeeze against institutional investors. This led to extreme price fluctuations and volatility, highlighting how coordinated actions by a minority group can generate large-scale market disruptions. It underscores the relevance of the Minority Game in understanding contemporary financial phenomena.

In other context, El-Shagi and Von Schweinitz [49] applies an ABM to the virtual economy of Diablo 3, an online multiplayer game. The authors analyze economic phenomena such as price inflation, wealth inequality, and economic mobility, with a focus on the heterogeneous price development of in-game commodities. The model successfully reproduces key features of Diablo 3's economy, such as the high and unequal prices of rare items and the difficulty for less wealthy players to move up economically. The study provides insights into how virtual economies can serve as simplified models for understanding real-world economic issues, especially regarding the effects of monetary policies and wealth distribution.

These developments from modeling market microstructures and wealth dynamics to exploring strategic interactions in competitive environments highlight the ever growing complexity and applicability of ABMs provide deep insights into market mechanics and economic distribution patterns, capturing the emergent behaviors arising from individual interactions. They offer a powerful framework for analyzing and predicting economic phenomena, informing both theoretical research and practical policy making.

5. Conclusions

In this chapter, we have explored the multifaceted applications of Agent-Based Models (ABMs) across various domains, including socio-physics, biology and economy. Beginning with the foundational principles of ABMs, we discussed how these models simulate complex systems through the interactions of autonomous agents, each with specific characteristics and behavioral rules. The integration of dynamic systems theory provided a mathematical framework to understand how micro-level interactions evolve into macro-level patterns, enhancing our comprehension of complex phenomena.

In socio-physics, we examined models like the Sznajd model for opinion dynamics, which illustrates how local interactions can lead to consensus or polarization within a community. We also explored Epstein's civil violence model, highlighting the factors that contribute to social unrest and the effectiveness of interventions. The application of ABMs to urban transportation policies demonstrated how these models can inform sustainable and equitable transportation strategies by simulating commuter behavior under various policy scenarios.

In the context of biological systems, we delved into how ABMs simulate the behavior of magnetic nanoparticles in medical treatments like magnetic fluid hyperthermia. This application showcased the ability of ABMs to model complex biological interactions at the micro-level, providing insights that can improve therapeutic outcomes.

The economics section highlighted the versatility of ABMs in modeling market behaviors, wealth distribution, and strategic interactions such as the Minority Game. By capturing the intricacies of financial markets and wealth dynamics, these models offer valuable perspectives on issues like market manipulation, wealth inequality, and the impact of individual strategies on market stability.

In education, we discussed how ABMs can model teaching-learning processes, capturing the dynamic interactions between students and educators. These models help in understanding how different pedagogical approaches and group dynamics influence knowledge acquisition and student performance, providing quantitative tools for curriculum design and educational policy.

Overall, agent-based models serve as powerful tools for analyzing and understanding complex systems across various fields. By simulating the interactions of individual agents within their environments, ABMs enable researchers and policymakers to predict emergent behaviors, design effective interventions, and foster systems that are more efficient, equitable, and sustainable.

Acknowledgements

We would like to thank Proyecto UTA MAYOR code 4198 – 4728-23, ANID BECAS/DOCTORADO NACIONAL Folio:21211479 and PROYECTO ANID VIU code: VIU23P0093. Additionally, we extend our gratitude to Ignacio Ormazabal, Ivana Stankov, Katarzyna Weron, and Makram El-Shagi for generously providing copyright permission for their figures.

Author details

Boris Atenas[1,2†*], Edward Larroza[2†] and Verónica Bahoz[3]

1 Universidad de Tarapacá, Arica, Chile

2 Universidad Católica del Norte, Antofagasta, Chile

3 Universidad de Antofagasta, Antofagasta, Chile

*Address all correspondence to: batenas@academicos.uta.cl

†These authors contributed equally

References

[1] Ising E. Beitrag zur Theorie des Ferromagnetismus. Zeitschrift für Physik. 1925;**31**:253-258. DOI: 10.1007/BF02980577 [Accessed: September 6, 2024]

[2] Galam S, Gefen Y, Shapir Y. Sociophysics: A new approach for sociological collective behavior. I. Mean-behavior description of a strike. Journal of Mathematical Sociology. 1982;**9**:1-13

[3] Potts RB. Some generalized order-disorder transformations. Mathematical Proceedings of the Cambridge Philosophical Society. 1952;**48**(1):106-109. DOI: 10.1017/S0305004100027419 [Accessed: September 6, 2024]

[4] Sznajd-Weron K, Sznajd J. Opinion evolution in closed community. International Journal of Modern Physics C. 2000;**11**(6):1157-1165. DOI: 10.1142/S0129183100000936 [Accessed: September 6, 2024]

[5] Nyczka P, Sznajd-Weron K. Anticonformity or Independence?—Insights from statistical physics. Journal of Statistical Physics. 2013;**151**:174-202. DOI: 10.1007/s10955-013-0701-4 [Accessed: September 6, 2024]

[6] Sznajd-Weron K, Sznajd J, Weron T. A review on the Sznajd model—20 years after. Physica A: Statistical Mechanics and Its Applications. 2021;**565**:125537. DOI: 10.1016/j.physa.2020.125537 [Accessed: September 6, 2024]

[7] Muslim R, Kholili MJ, Nugraha ART. Opinion dynamics involving contrarian and independence behaviors based on the Sznajd model with two-two and three-one agent interactions. Physica D: Nonlinear Phenomena. 2022;**439**:133379. DOI: 10.1016/j.physd.2022.133379 [Accessed: September 6, 2024]

[8] Azhari et al. Independence role in the generalized Sznajd model. Physica A: Statistical Mechanics and Its Applications. 2024;**652**:130042. DOI: 10.1016/j.physa.2023.130042 [Accessed: September 6, 2024]

[9] Burbach L, Halbach P, Ziefle M, Calero Valdez A. Who shares fake news in online social networks?: An agent-based model of different personality models and behaviors in social networks. Proceedings of the. In: 27th Conference on User Modeling, Adaptation and Personalization. New York, NY, USA: Association for Computing Machinery (ACM); 2019. pp. 234-243. DOI: 10.1145/3320435.3320456 [Accessed: August 20, 2024]

[10] Zhao X, Wang S, Chen L. The structural dynamics of fake news dissemination in social media networks: An empirical study. Social Network Analysis and Mining. 2023;**12**(2):101-119. DOI: 10.1007/s13278-023-01028 [Accessed: August 20, 2024]

[11] Epstein JM. Modeling civil violence: An agent-based computational approach. Proceedings of the National Academy of Sciences. 2002;**99**(Suppl 3):7243-7250. DOI: 10.1073/pnas.092080199 [Accessed: August 20, 2024]

[12] Lemos C, Lopes RJ, Coelho H. Analysis of the decision rule in Epstein's agent-based model of civil violence. In: 2015 Third World Conference on Complex Systems (WCCS). Amsterdam, Netherlands: Elsevier; 2015. DOI: 10.1109/ICoCS.2015.7483303 [Accessed: August 20, 2024]

[13] Ormazábal I, Borotto FA, Astudillo HF. Influence of money distribution on civil violence model. Complexity. 2017;**2017**:1-15. DOI: 10.1155/2017/7453560 [Accessed: August 8, 2024]

[14] Ormazábal I, Urbina F, Borotto FA, Astudillo HF. Phase diagram in a one-dimensional civil disorder model. Physical Review E. 2022;**105**:054110. DOI: 10.1103/PhysRevE.105.054110 [Accessed: August 8, 2024]

[15] Maggi E, Vallino E. Understanding urban mobility and the impact of public policies: The role of the agent-based models. Research in Transportation Economics. 2016;**63**:1-10. DOI: 10.1016/j.retrec.2016.04.010 [Accessed: August 20, 2024]

[16] Zhuge C, Wei B, Dong C, Shao C, Shan Y. Exploring the future electric vehicle market and its impacts with an agent-based spatial integrated framework: A case study of Beijing, China. Journal of Cleaner Production. 2019;**221**:710-737. DOI: 10.1016/j.jclepro.2019.02.262 [Accessed: August 10, 2024]

[17] Wise S, Crooks A, Batty M. El transporte en el modelado urbano basado en agentes. In: Namazi-Rad M-R et al., editors. *Modelado basado en agentes de sistemas urbanos*. Springer; 2017. pp. 129-148. DOI: 10.1007/978-3-319-51957-9_8 [Accessed: August 20, 2024]

[18] Stankov et al. Uncovering physical activity trade-offs in transportation policy: A spatial agent-based model of Bogotá, Colombia. International Journal of Behavioral Nutrition and Physical Activity (London, United Kingdom: BioMed Central). 2024;**21**:54. DOI: 10.1186/s12966-024-01570-1

[19] Achachlouei MA, Hilty LM. System dynamics vs. agent-based modeling—Comparing models and approaches: A literature review and a transformation procedure. Journal of Environmental Modelling & Software. KTH Royal Institute of Technology, Division of Environmental Strategies Research, Stockholm, Sweden. 2015:1-25

[20] Bordogna CM, Albano EV. Theoretical description of teaching–learning processes: A multidisciplinary approach. Physical Review Letters. 2001;**87**:118701. DOI: 10.1103/PhysRevLett.87.118701 [Accessed: September 6, 2024]

[21] Gu X, Blackmore KL. A systematic review of agent-based modelling and simulation applications in the higher education domain. Higher Education Research & Development. 2015;**34**(5):883-898. DOI: 10.1080/07294360.2015.1011088 [Accessed: September 11, 2024]

[22] Ormazábal I, Borotto FA, Astudillo HF. An agent-based model for teaching–learning processes. Physica A. 2021;**565**:125563. DOI: 10.1016/j.physa.2020.125563 [Accessed: September 11, 2024]

[23] Velázquez L, Atenas B, Castro-Palacio JC. Quantitative methods to determine the student workload I: An empirical study case based on digital platforms. Chaos. 2022;**32**:000000. DOI: 10.1063/5.0103719 [Accessed: August 8, 2024]

[24] Atenas B, Velázquez L, Castro-Palacio JC. Quantitative methods to determine the student workload II: Statistical models for the microcurricular performance indicators. Chaos. 2022;**32**:000000. DOI: 10.1063/5.0104307 [Accessed: August 8, 2024]

[25] Velazquez L, Atenas B, Castro-Palacio JC. An adaptive methodology for curriculum redesign based on performance indicators of the student progression. Studies in Higher Education. 2024;**49**(3):410-440. DOI: 10.1080/03075079.2023.2238751 [Accessed: August 8, 2024]

[26] Pleyer J, Fleck C. Agent-based models in cellular systems. Frontiers in Physics. 2023;**10**:968409. DOI: 10.3389/fphy.2022.968409 [Accessed: September 25, 2024]

[27] Wei H, Feng C, Li F. Modeling biological memory network by an autonomous and adaptive multi-agent system. Brain Informatics. 2024;**11**:23. DOI: 10.1186/s40708-024-00237-8

[28] Schinko C et al. Accelerated airborne virus spread simulation: Coupling agent-based modeling with GPU-accelerated computational fluid dynamics. In: Proceedings of the 17th International Joint Conference on Computer Vision, Imaging and Computer Graphics Theory and Applications (VISIGRAPP 2022). GRAPP. Setúbal, Portugal: SCITEPRESS – Science and Technology Publications, Lda.; 2022. pp. 278-285. DOI: 10.5220/0010904500003124

[29] Fernandes RS, Vivas Miranda JG. An agent-based model for studying the temperature changes on environments exposed to magnetic fluid hyperthermia. Computers in Biology and Medicine. 2024;**170**:108053. DOI: 10.1016/j.compbiomed.2024.108053 [Accessed: September 6, 2024]

[30] Fortin JP, Gazeau F, Wilhelm C. Intracellular heating of living cells through Néel relaxation of magnetic nanoparticles. European Biophysics Journal. 2008;**37**(2):223-228. DOI: 10.1007/s00249-007-0197-4 [Accessed: September 6, 2024]

[31] Murray JD. Mathematical Biology: I. An Introduction, Vol. 17 of Interdisciplinary Applied Mathematics. New York, NY, USA: Springer; 2002

[32] Willis G. Income distribution and income shares: Wealth and income distributions explained using generalised Lotka-Volterra SFC ABM models. International Review of Applied Economics. 2015;**29**(6):816-842. DOI: 10.1080/02692171.2015.1065225 [Accessed: September 25, 2024]

[33] Palombi F, Ferriani S, Toti S. Coevolutionary dynamics of a variant of the cyclic Lotka-Volterra model with three-agent interactions. The European Physical Journal B. 2020;**93**(194):1-18. DOI: 10.1140/epjb/e2020-100552-5 [Accessed: September 25, 2024]

[34] Chakraborti A, Muni Toke I, Patriarca M, Abergel F. Econophysics review: II. Agent-based models. Quantitative Finance. 2011;**11**(7):1013-1041. DOI: 10.1080/14697688.2010.539249 [Accessed: September 7, 2024]

[35] Bouchaud JP, Mézard M. Wealth condensation in a simple model of economy. Physica A: Statistical Mechanics and Its Applications. 2000;**282**(3-4):536-545. DOI: 10.1016/S0378-4371(00)00205-3 [Accessed: September 7, 2024]

[36] Lux T, Marchesi M. Volatility clustering in financial markets: A microsimulation of interacting agents. International Journal of Theoretical and Applied Finance. 2000;**3**(4):675-702. DOI: 10.1142/S0219024900000826 [Accessed: September 7, 2024]

[37] Chiarella C, Iori G. A simulation analysis of the microstructure of double auction markets. Journal of Physics: Conference Series. 2002;**2**(5):303-313.

DOI: 10.1088/1469-7688/2/5/303 [Accessed: September 18, 2024]

[38] Mike S, Farmer JD. An empirical behavioral model of liquidity and volatility. Journal of Economic Dynamics and Control. 2008;**32**:200-234. DOI: 10.1016/j.jedc.2007.01.025 [Accessed: September 18, 2024]

[39] Dragulescu AA, Yakovenko VM. Statistical mechanics of money. The European Physical Journal B. 2000;**17**(4):723-729. DOI: 10.1007/s100510070114 [Accessed: September 10, 2024]

[40] Chakraborti A, Chakrabarti BK. Statistical mechanics of money: How saving propensity affects its distribution. The European Physical Journal B. 2000;**17**(1):167-170. DOI: 10.1007/s100510070173 [Accessed: September 10, 2024]

[41] Mohanty PK. Generic features of the wealth distribution in ideal-gas-like markets. Physical Review E. 2006;**74**(1):011117. DOI: 10.1103/PhysRevE.74.011117 [Accessed: September 10, 2024]

[42] Aydiner E, Cherstvy AG, Metzler R, Sokolov IM. Universal Pareto laws in agent-based exchange models: Debt and varying initial-money distributions. The European Physical Journal B. 2023;**96**:123. DOI: 10.1140/epjb/s10051-023-00579-y [Accessed: September 18, 2024]

[43] Fratrič P, Sileno G, Klous S, van Engers T. Manipulation of the bitcoin market: An agent-based study. Financial Innovation. 2022;**8**:53. DOI: 10.1186/s40854-022-00364-3 [Accessed: September 18, 2024]

[44] Arthur WB. Inductive reasoning and bounded rationality. The American Economic Review. 1994;**84**(2):406-411. Available from: https://www.jstor.org/stable/2117868 [Accessed: August 23, 2024]

[45] Challet D, Zhang YC. Emergence of cooperation and organization in an evolutionary game. Physica A: Statistical Mechanics and Its Applications. 1997;**246**(3-4):407-418. DOI: 10.1016/S0378-4371(97)00419-6 [Accessed: August 23, 2024]

[46] Challet D, Zhang YC. On the minority game: Analytical and numerical studies. Physica A: Statistical Mechanics and Its Applications. 1998;**256**(3-4):514-532. DOI: 10.1016/S0378-4371(98)00260-X [Accessed: August 23, 2024]

[47] Zhang XJ et al. Dynamics of cooperation in minority games in alliance networks. Scientific Reports. 2018;**8**(1):4052. DOI: 10.1038/s41598-018-22361-7 [Accessed: August 9, 2024]

[48] Xu WJ, Zhong LX. Evolutionary dynamics in financial markets with heterogeneities in investment strategies and reference points. PLoS One. 2022;**17**(7):e0288277. DOI: 10.1371/journal.pone.0288277 [Accessed: August 9, 2024]

[49] El-Shagi M, von Schweinitz G. The diablo 3 economy: An agent based approach. Computational Economics. 2016;**47**(2):193-217. DOI: 10.1007/s10614-014-9480-5 [Accessed: August 9, 2024]

Chapter 6

Chaotic Dynamics Derived from the Montgomery Conjecture: Application to Electrical Systems

Zeraoulia Rafik, Alvaro Humberto Salas and Ayadi Souad

Abstract

Here, we introduce a novel method for obtaining chaotic dynamics based on the Montgomery conjecture for the pair correlation of zeros of the Riemann zeta function. Motivated by the conjecture, we present a recursive relation that reveals chaotic behavior. Notably, we provide insights into the possible uses of this derived chaotic dynamics in electrical engineering by interpreting it as a unique representation of an electrical system. Furthermore, we investigate the relevance of entropy, bifurcation analysis, and chaos theory in this framework for electrical systems. We look into its applicability to signal processing, stability analysis through bifurcation, and how entropy measures the predictability or unpredictability of electrical signals. Additionally, we discuss the system's strange attractor and its transition to voltage collapse, highlighting the interplay between chaotic dynamics and stability in electrical systems. Furthermore, we analyze the system's energy distribution, taking into account how chaotic dynamics may affect energy allocation or dissipation. Furthermore, we compare the chaotification and Hermiticity of the resulting operators between Yitang dynamics and Montgomery dynamics. To have a better grasp of the spectrum features of each operator, we calculate the eigenvalues for each one obtained from the corresponding dynamics. Our results provide fresh insights into number-theoretic chaotic dynamics and how they might be applied in real-world electrical engineering applications. This work provides encouraging opportunities for further research and technology developments by laying the foundation for creative investigations in system dynamics.

Keywords: Montgomery's Pair Correlation Conjecture, chaotic dynamics, control theory, energy storage, chaotic operator

1. Introduction

Number theory, the study of the properties and relationships of integers, has long been a fertile ground for deep conjectures that continue to intrigue mathematicians. These conjectures often emerge from patterns observed in numerical data, leading to hypotheses whose proof could significantly advance our understanding of fundamental mathematical principles.

A major area of interest in number theory is the distribution of prime numbers. Famous conjectures in this field include the Twin Prime Conjecture, Goldbach's

Conjecture, and the Riemann Hypothesis. Despite their deceptively simple formulations, these conjectures have resisted proof for centuries, captivating mathematicians and driving numerous research efforts.

In recent years, there has been a growing interest in the potential links between number theory and chaotic dynamics. Chaos theory, which studies complex and unpredictable behavior in deterministic systems, has found diverse applications across fields like physics, biology, and economics. The possibility that chaotic behavior could emerge from deterministic number-theoretic systems has inspired investigations into these unexpected connections [1].

One promising line of research explores how chaotic dynamics can be applied to conjectures in number theory. By analyzing chaotic systems inspired by conjectures such as the Riemann Hypothesis and the Twin Prime Conjecture, researchers seek to uncover new structural insights that could eventually lead to proofs or disproofs.

In this work, we build on our earlier research presented in Ref. [2], where we introduced a novel derivation of chaotic dynamics based on the Montgomery Pair Correlation Conjecture. This conjecture, which describes statistical properties of the non-trivial zeros of the Riemann zeta function [3], provides a fertile framework for studying the interplay between number theory and chaotic behavior.

Building upon our previous findings, we offer an enhanced analysis and a more comprehensive exploration of chaotic dynamics derived from the Montgomery conjecture. Our investigation not only deepens the understanding of these dynamics but also highlights their potential applications in electrical engineering.

Several recent works have explored advanced control strategies and mathematical techniques that resonate with our study. For instance, Deep et al. [4] introduced a robust moment-matching load frequency control strategy for cyber-physical power systems, addressing communication delays, which has implications for stability analysis. Similarly, Ansari and Raja [5] developed an enhanced cascaded frequency controller optimized by a flow direction algorithm, demonstrating the importance of control optimization in complex dynamical systems. Aryan et al. [6] proposed an equilibrium optimizer-tuned frequency-shifted internal model control design for industrial plants, emphasizing structured control in practical applications.

Mukherjee et al. [7] investigated fractional backstepping and Lyapunov strategies for stabilization, which align with the stability concerns inherent in chaotic systems. Kumari et al. [8] explored fuzzy control methods for frequency regulation in power systems, demonstrating the relevance of advanced control techniques in managing nonlinear dynamics. Further, Hu et al. [9] introduced a double predictive proportional-integral control strategy, relevant for handling delays and nonlinearities similar to those found in chaotic number-theoretic models.

Mehta et al. [10] developed a tri-parametric fractional-order controller for integrating systems with time delays, a topic closely linked to our interest in delay-induced chaotic behavior. Mukherjee et al. [11] proposed an optimal fractional Lyapunov-based adaptive control scheme, which has implications for the stability of dynamical systems derived from number-theoretic conjectures. Aryan and Raja [12] extended this line of research with an LFC scheme incorporating renewables and electric vehicle penetration, further highlighting the intersection between applied engineering and mathematical modeling.

Other notable contributions include Aryan et al. [7], who examined a RIMC-based dual-loop strategy for unstable processes, and their work on internal model control for boiler steam drum regulation [13], both of which provide insights into structured control methodologies applicable to our chaotic framework. Additionally,

Anand et al. [14] investigated type-2 fuzzy controllers optimized *via* equilibrium algorithms, reinforcing the role of computational optimization in nonlinear control systems.

By drawing parallels between number theory and electrical systems, we uncover new insights that contribute to both fields. This interdisciplinary approach seeks to bridge the gap between chaos theory, number theory, and applied engineering sciences [15]. We aim to foster new avenues for research and innovation at the intersection of these domains [16].

2. Notations used

In this section, we provide a list of notations and symbols used throughout the chapter for the reader's convenience.

- $F(\omega)$: The function representing the chaotic dynamics in the system.
- $\delta(\omega)$: The Dirac delta function, representing an impulse at $\omega = 0$.
- C: A constant used in the formulation of the chaotic system.
- $\text{sgn}(\omega)$: The sign function, defined as:

$$\text{sgn}(\omega) = \begin{pmatrix} 1 & \text{if } \omega > 0 \\ 0 & \text{if } \omega = 0 \\ -1 & \text{if } \omega < 0 \end{pmatrix}$$

- $\text{sgn}'(\omega)$: The first derivative of the sign function with respect to ω.
- $\text{sgn}''(\omega)$: The second derivative of the sign function with respect to ω.
- r: A parameter related to the chaotic dynamics.
- ${}_0F_1(;b;z)$: The confluent hypergeometric function.
- ${}_0\tilde{F}_1(;b;z)$: A variant of the confluent hypergeometric function used in the context of chaotic systems.
- i: The imaginary unit, defined as $i = \sqrt{-1}$.
- ω: The frequency variable in the context of the chaotic dynamics.

2.1 Main results

2.1.1 Derived electrical system

The chaotic dynamics inspired by Montgomery's Pair Correlation Conjecture are represented by the following recursive relation, interpreted as an electrical system:

$$x_{n+1} = 1 - \left(\frac{\sin(\pi/x_n)}{r\pi/x_n}\right)^2 + \frac{R}{L} + \frac{1}{Cx_n}$$

where x_n represents the system state, and R, C, L, and r denote the electrical parameters (resistance, capacitance, inductance, and an additional scaling parameter, respectively).

2.1.2 Parameter analysis insights

An investigation of how different electrical parameters influence system behavior revealed the following:

- *Inductance dominance* ($L > R$): Leads to smoother transitions and stability in the system.
- *Capacitance and inductance*: Large values of C and L tend to produce periodic behavior, with regular oscillations in the system.
- *Sensitivity to parameter r*: Small changes in r cause significant shifts in system behavior, including the appearance of chaotic dynamics.

2.1.3 System dynamics and analysis

The derived electrical system demonstrates a range of dynamic behaviors depending on parameter values. Key findings include:

- *Phase portraits*: Numerical simulations show stable fixed points when $r = 0.7$, where the system tends to settle into predictable patterns.
- *Limit cycles and frequency analysis*: Fourier analysis reveals the system's dominant frequencies, corresponding to periodic behavior under certain parameter conditions.
- *Poincaré section*: Recurrent states appear in the negative region, highlighting the chaotic nature of the system.

2.1.4 Quantum system interpretation

By interpreting the electrical system as a quantum system, the Schrödinger equation with a potential function $V(x) = A + \frac{1}{Cx}$ yields sinusoidal eigenfunctions. These eigenfunctions correspond to discrete energy levels and bound states, indicating parallels between quantum mechanics and signal processing. The eigenfunctions represent fundamental frequency components of the electrical system, essential for stability and performance analysis.

2.2 Detailed energy dynamics analysis

2.2.1 Influence of parameter r

The parameter r plays a crucial role in determining energy storage and distribution in the system. Specifically:

- *Energy storage sensitivity*: As r varies, significant shifts in energy storage behavior occur, affecting both system stability and energy distribution.
- *Energy variation with capacitance* (C): The system shows heightened energy storage in the inductor as C decreases, suggesting a direct relationship between capacitance and energy dynamics.

2.2.2 Energy distribution implications

The context of the derived system is represented by the equation:

$$f(x, r, R, L, C) = 1 - \left(\frac{\sin\left(\frac{\pi}{x}\right)}{r\pi / x} \right)^2 + \frac{R}{L} + \frac{1}{Cx}$$

with parameters $R = 0.000025$, $L = 0.00045$, and $C = 0.73$, and the variation of r has a significant effect on energy storage. Understanding the dynamics of r could reveal optimal ranges for energy efficiency and system stability.

2.3 Eigenvalue and operator analysis

2.3.1 Comparison of eigenvalues

- *Yitang dynamics*: The first eigenvalue of the Yitang system, computed using the Implicitly Restarted Arnoldi Method (IRAM), is $7.11343 - 0.178929i$, indicating complex eigenvalues and potential chaotic behavior.
- *Montgomery dynamics*: The first eigenvalue of the Montgomery system is -2.99007, a real value, suggesting more regular, predictable dynamics.

2.3.2 Diagonalizability

- *Yitang dynamics*: The complex eigenvalues indicate the possibility of complex eigenvectors, and further analysis is required to determine whether the system is diagonalizable.
- *Montgomery dynamics*: The real eigenvalues suggest that the Montgomery system's corresponding operator is diagonalizable, with real and linearly independent eigenvectors.

3. Montgomery's Pair Correlation Conjecture in number theory

Montgomery's Pair Correlation Conjecture is a significant hypothesis in number theory that explores the statistical distribution of spacings between the non-trivial zeros of the Riemann zeta function. Proposed by Hugh L. Montgomery, this conjecture has established deep connections between the distribution of these spacings and random matrix theory [17].

3.1 Background

The Riemann zeta function, $\zeta(s)$, is a complex-valued function of a complex variable s, whose non-trivial zeros are central to understanding the distribution of prime numbers. These non-trivial zeros, also referred to as the Riemann zeros, lie in the critical strip where the real part of s is between 0 and 1, and play a fundamental role in number theory. Montgomery's conjecture focuses on the statistical properties of the gaps between these zeros.

3.2 The pair correlation conjecture

Montgomery's Pair Correlation Conjecture predicts that the pair correlation function $R_2(\Delta)$, which measures the probability density of finding two non-trivial zeros with a normalized spacing Δ, converges to a distribution analogous to the eigenvalue spacings of matrices in the Gaussian Unitary Ensemble (GUE) as the imaginary part of the zeros grows large. Specifically, the conjecture can be formulated as:

$$R_2(\Delta) \to \text{limiting distribution of GUE}, \quad \text{as} \quad \operatorname{Im}(\rho) \to \infty, \tag{1}$$

where ρ denotes the non-trivial zeros of the Riemann zeta function, and GUE refers to a class of random matrices whose eigenvalue statistics are well-understood through random matrix theory.

3.3 The pair correlation function

The pair correlation function $R_2(\Delta)$ is mathematically defined as:

$$R_2(\Delta) = \lim_{\substack{T\to\infty \\ T \text{ fixed}}} \frac{1}{T} \sum_{\substack{0<\operatorname{Im}(\rho_j),\, \operatorname{Im}(\rho_k)<T \\ |\operatorname{Im}(\rho_j)-\operatorname{Im}(\rho_k)|<\Delta}} 1, \tag{2}$$

where ρ_j and ρ_k are the non-trivial zeros of the Riemann zeta function, and Δ represents the normalized spacing between their imaginary parts.

3.4 Pair correlation formula with sine function

Montgomery's conjecture can also be represented by a pair correlation formula involving the sine function, which approximates the behavior of the zeros. The formula is given by:

$$R_2(\Delta) = \frac{2}{\pi}\left(\Delta \sin\left(\frac{\pi\Delta}{2}\right) + \frac{\sin^2\left(\frac{\pi\Delta}{2}\right)}{2}\right), \tag{3}$$

where Δ denotes the normalized spacing between the non-trivial zeros. This formula highlights the oscillatory nature of the pair correlation function and its dependence on Δ.

3.5 Implications and significance

The significance of Montgomery's Pair Correlation Conjecture lies in its surprising connection between number theory and random matrix theory. The conjecture suggests that the statistical behavior of the non-trivial zeros of the Riemann zeta function mirrors the eigenvalue distribution of random matrices from the GUE. This connection supports the universality phenomenon observed in random matrix theory and deepens our understanding of both the analytic and probabilistic aspects of number theory. Verification of this conjecture would not only strengthen these connections but also provide further insight into the zeros of the Riemann zeta function and their role in the distribution of primes [18].

3.6 Current status

Montgomery's Pair Correlation Conjecture has received substantial numerical support over a wide range of zeros of the Riemann zeta function. Computational experiments have shown that the statistical properties of the gaps between the non-trivial zeros exhibit strong alignment with the predictions of the conjecture. However, despite these compelling numerical results, a rigorous mathematical proof remains elusive. Ongoing research aims to bridge this gap by delving deeper into the distribution of Riemann zeros and its broader implications in number theory, random matrix theory, and even fields like quantum chaos.

This research builds on Montgomery's pioneering work by seeking a more refined understanding of the underlying structures that govern the behavior of these zeros. One such avenue is to interpret these statistical models as dynamic systems, allowing for a novel exploration of the conjecture through the lens of dynamical theory.

3.7 Deriving the dynamic system

Our approach involves deriving a dynamic system inspired by Montgomery's Pair Correlation Conjecture [19]. The conjecture, which provides a statistical framework for understanding the distribution of non-trivial zeros, can be reinterpreted as a dynamic model that encapsulates the behavior of these zeros over time. Specifically, the pair correlation function $g(u)$, a key feature of the conjecture, describes the probability distribution of normalized spacings between zeros. It is typically represented as:

$$g(u) = 1 - \left(\frac{\sin(\pi u)}{\pi u}\right)^2 + \delta(u),$$

where $\delta(u)$ accounts for deviations from the idealized statistical behavior. To explore the conjecture's implications further, we transform this pair correlation function into a recurrence relation that governs the evolution of a system state, x_n, at each iteration n.

This transformation provides a dynamic interpretation of the conjecture, where the behavior of the zeros of the Riemann zeta function is mirrored in the evolution of the system's state. Our derived dynamic system takes the form of the following recurrence relation:

$$x_{n+1} = 1 - \left(\frac{\sin(\pi/x_n)}{r\pi/x_n}\right)^2 + \frac{R}{L} + \frac{1}{Cx_n},$$

where x_n represents the state of the system at iteration n, and the parameters R, C, L, and r correspond to resistance, capacitance, inductance, and external influences, respectively, analogous to those found in electrical circuits. This formulation not only draws a striking parallel between number theory and electrical engineering but also opens the door to the exploration of chaotic dynamics in an electrical context.

By formulating the conjecture in terms of a dynamic system, we can investigate the evolution of zeros as a complex system evolving over time, shedding light on potential chaotic behavior, bifurcations, and stability phenomena. This connection offers new tools to analyze the conjecture's deeper implications and to possibly uncover patterns that may bring us closer to a formal proof of the conjecture itself [20–22].

4. Response and robustness of the new electrical system

This section delves into the response and robustness analysis of our newly derived electrical system, exploring its behavior under specific parameter values. Understanding how variations in component values influence the system's dynamics is crucial for our exploration of the conjecture's implications. By examining the stability and response of this dynamic system, we aim to uncover insights that may parallel the statistical properties predicted by Montgomery's Pair Correlation Conjecture, ultimately illuminating the underlying structure of Riemann zeros [23].

4.1 Effect of larger L compared to R in the electrical system

The corrected error term in the Montgomery conjecture, $\delta(u)$, embodies a small deviation, suggesting that inductance (L) surpasses resistance (R) in magnitude. This relationship is significant, as it directly impacts the behavior of our dynamic system, reflecting the conjecture's subtle corrections. In our analysis, we set $\delta(u) = \frac{R}{L} + \frac{1}{Cx}$ with $R = 10$, $C = 0.1$, and $L = 20$, allowing us to explore the influence of varying r values on system dynamics.

The plot in **Figure 1** illustrates how the electrical system responds when L is larger than R, emphasizing the importance of inductance in shaping the system's dynamics. The yellow and blue fluctuations represent distinct responses under different r values, showcasing how the system adapts and changes in behavior as the parameter r varies [19, 24–26].

- *Influence of larger L*: When L dominates R, the system's behavior is significantly affected by inductance, highlighting altered dynamics compared to scenarios where R is more dominant. This suggests that in systems with characteristics analogous to Riemann zeros, the inductive influences may play a more prominent role in shaping zero distributions.

- *Reduced R/L contribution*: The R/L ratio diminishes due to the increased inductance, which can lead to smoother transitions or more controlled oscillations. This reduction emphasizes the nuanced interplay between resistance

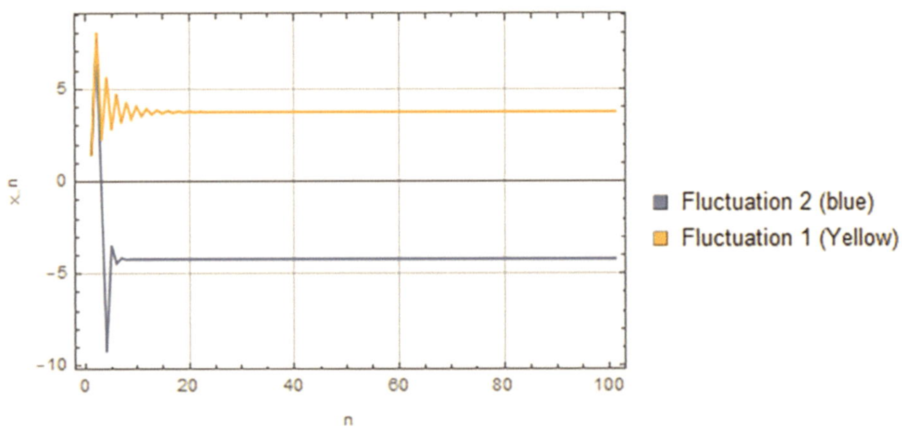

Figure 1.
Behavior of the electrical system under L > R influence for R = 10, C = 0.1, L = 20.

and inductance, mirroring how minor deviations in the conjecture may affect the overall statistical behavior of zeros.

- *Roles of C and R*: Although L is the prevailing factor, capacitance (C) and resistance (R) continue to exert influence. Capacitance governs the charging and discharging behavior, akin to how the distribution of zeros may fluctuate under different theoretical models. Resistance contributes to the overall damping of the system, paralleling how statistical noise might obscure or reveal underlying patterns in the distribution of Riemann zeros.

4.2 Periodic behavior with larger C and L dominance

In this scenario, we explore the system's behavior with significantly larger capacitance ($C = 50$) while maintaining a smaller resistance ($R = 10$) compared to the inductance ($L = 20$). This configuration emphasizes how an increase in capacitance, when coupled with a dominant inductance, can lead to distinct periodic responses in the electrical system. The resulting plot, as shown in **Figure 2**, illustrates this periodic response under varying r values.

- *Observation of periodicity*: The blue trace in the plot exhibits distinct periodic fluctuations, indicating a recurring pattern in the system's response as r varies. This periodicity suggests that the larger capacitance plays a crucial role in storing and releasing energy, creating oscillations that are influenced by the inductance. In contrast, the yellow trace demonstrates a stable curve, indicative of consistent and non-periodic behavior across the iterations.

- *Influence of* C and L: The dominance of L over R in conjunction with the increased C establishes conditions conducive to oscillatory dynamics. The periodic response reflected in the blue trace signifies how the capacitance enhances the system's ability to oscillate, as it can accumulate and release energy more effectively, emphasizing the interplay between these two parameters.

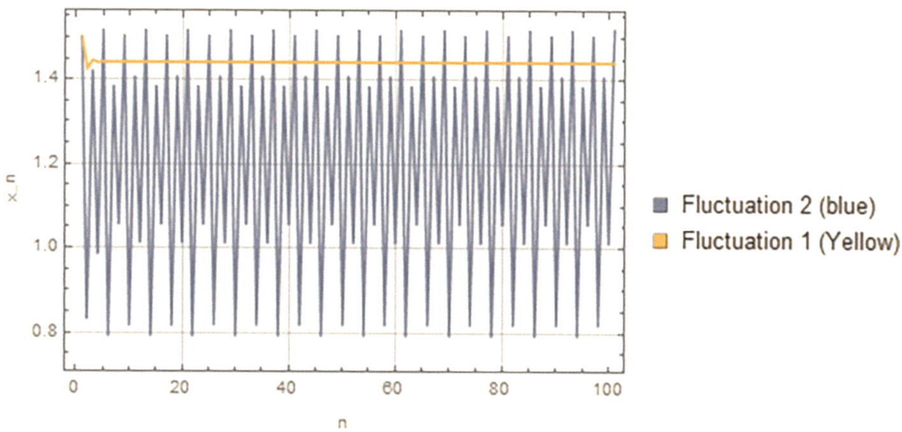

Figure 2.
Periodic behavior of the electrical system under larger C and dominant L for $R = 10$, $C = 50$, $L = 20$.

- *Stability and regularity*: While the blue trace shows periodic behavior, the yellow trace remains stable, demonstrating that under specific configurations, the system can exhibit both dynamic and static responses. This duality could mirror the complexity seen in the statistical behaviors predicted by conjectures like Montgomery's, where different configurations of parameters lead to different statistical properties in the distribution of Riemann zeros.

4.3 Sensitivity to *r* values and absence of yellow fluctuation

The observed absence of the yellow fluctuation in the plot under specific ranges of *r* values indicates a sensitivity of the electrical system to variations in *r*. The plot in **Figure 3** highlights this sensitivity, particularly within a certain range of *r* values (between 0.008 and 0.07 in increments of 0.7), leading to a significant change in the system's behavior.

• *Effect of r sensitivity*: Small changes in *r* appear to induce significant alterations in the system's response, showcasing the system's sensitivity to this parameter. This highlights how specific configurations can lead to critical changes in behavior, similar to how slight perturbations in conjectures can drastically alter outcomes in number theory.

• *Disappearance of yellow fluctuation*: The absence of the yellow fluctuation within this *r* value range suggests a shift in the system's dynamics, where dominant behaviors may eclipse others. The elimination of the yellow trace indicates a transition to a new regime of system dynamics, possibly influenced by the increased capacitance and inductance.

• *Values of* x_n: Within this range of *r* values, the values of x_n become increasingly large and negative, indicating divergence or a drastic change in the system's behavior. This divergence could signify a critical transition or failure state in the system, reflecting the complex dynamics that can emerge from interactions among *C*, *L*, and *R*. Such transitions could draw parallels to critical states observed in statistical physics and number theory, where specific conditions lead to significant changes in behavior.

The plot depicted in **Figure 3** illustrates the system's sensitivity to *r* values, evidenced by the disappearance of the yellow fluctuation within the specified range. The divergence of x_n values into larger negative magnitudes signifies a substantial deviation from expected behavior, potentially indicating a critical state or failure

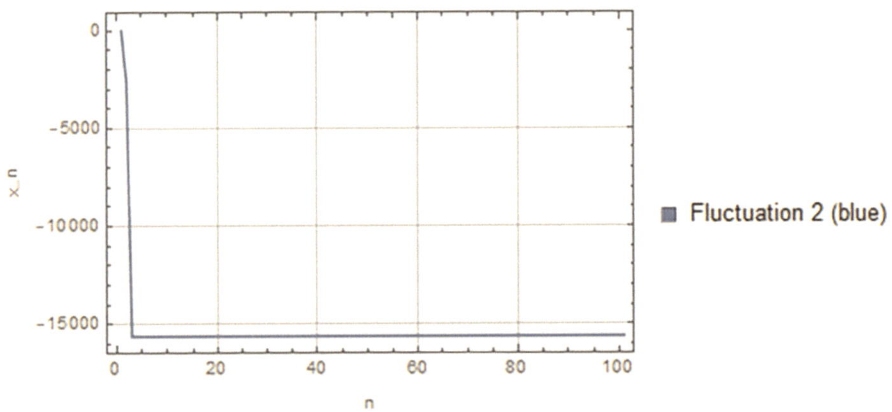

Figure 3.
Sensitivity of the electrical system to r values and absence of the yellow fluctuation for specified r range.

within the electrical system. This critical state reinforces the importance of understanding parameter interactions, as small variations in one parameter can lead to drastic changes, much like the intricate relationships explored in conjectures relating to the distribution of primes and the zeros of the Riemann zeta function [17, 18].

4.4 Comprehensive comparison of system responses to r variation

In this extended investigation, we analyze the system's behavior under different parameter settings by fixing $R = 0.8$, $C = 0.7$, and $L = 3$. The response of the system to varying r values within the range $0.5 \leq r \leq 2$ is illustrated in **Figure 4b**.

Figure 4 highlights the system's response for two distinct scenarios. The first scenario (left) showcases the absence of the yellow fluctuation under the parameters $R = 0.7$, $C = 0.3$, and $L = 0.5$. In contrast, the second scenario (right) displays the behavior of the system under the conditions $R = 0.8$, $C = 0.7$, and $L = 3$.

- Observation of fluctuations:
  - In **Scenario 1**, we notice stable output without significant fluctuations, indicating a balanced response of the system.
  - Conversely, in **Scenario 2**, the response exhibits larger fluctuations, potentially linked to the higher C and L values. This supports the earlier findings in the previous subsection regarding the influence of larger capacitance and inductance, which contribute to a more dynamic system behavior.

  Next, we delve into two additional scenarios where $R = 0.7$ and $C = 0.3$ while varying L within the ranges $L = 0.5$ (**Figure 5a**) and $L = 0.8$ (**Figure 5b**) for $0.5 \leq r \leq 2.5$.

Figure 5 compares the system's output across **Scenario 3** and **Scenario 4**. Here, both scenarios display a notable negative trend in the blue fluctuations, suggesting a decline in the system's output as L (increases)

- *Amplitude and stability analysis*:
  - In **Scenario 3**, with $L = 0.5$, the system displays significant oscillatory behavior, characterized by marked fluctuations in the blue trace.

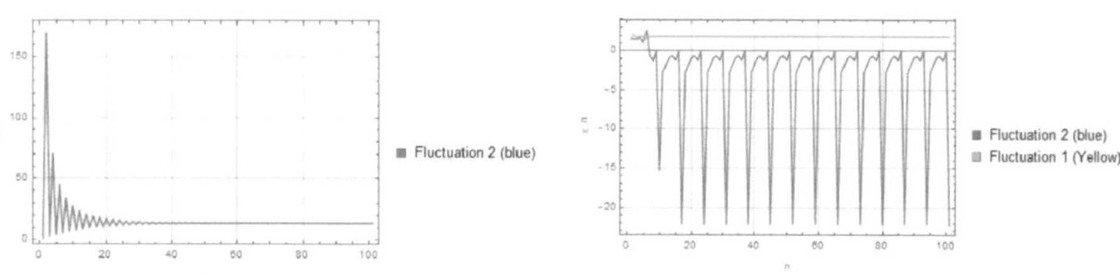

(a) (b)

Figure 4.
Comprehensive comparison of system responses. (a) Scenario 1: Absence of Yellow Fluctuation ($R = 0.7$, $C = 0.3$, $L = 0.5$) ($0.5 \leq r \leq 2$). (b) Scenario 2: System Response with $R = 0.8$, $C = 0.7$, and $L = 3$ ($0.5 \leq r \leq 2$).

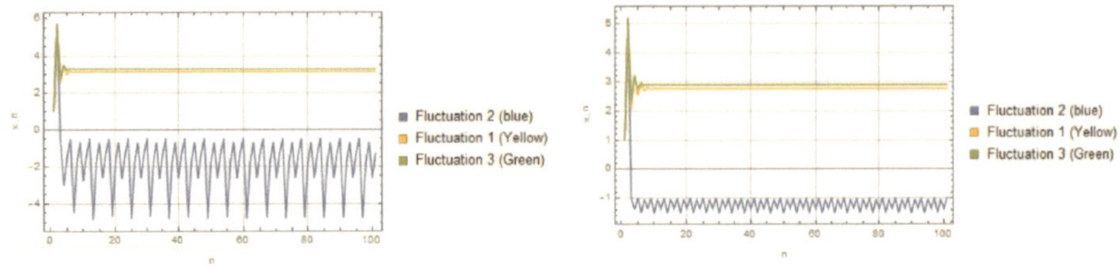

(a) (b)

Figure 5
Comprehensive comparison of system responses. (a) Scenario 3: $R = 0.7$, $C = 0.3$, $L = 0.5$ ($0.5 \leq r \leq 2.5$). (b) Scenario 4: $R = 0.7$, $C = 0.3$, $L = 0.8$ ($0.5 \leq r \leq 2.5$).

- However, in **Scenario 4**, when L is increased to 0.8, the amplitude of these fluctuations decreases considerably, indicating a transition toward a more stable regime. This behavior suggests a damping effect due to increased inductance, which could stabilize the oscillations and reduce the impact of fluctuations, reinforcing our previous observations about the influence of C and L in the electrical system.

- *Comprehensive insights*:

    - The combination of all explored scenarios reveals the intricate interplay between resistance, capacitance, and inductance in shaping the system's response to variations in r. The transitions from dynamic to stable behavior in response to parameter changes illustrate the complex dynamics at play, essential for understanding the operational characteristics of the electrical system.

In summary, this comprehensive comparison enriches our understanding of the system's response to varying conditions, emphasizing the significant impact of C and L on system dynamics, as well as the sensitive dependence on r values, consistent with the findings discussed in earlier subsections.

4.4.1 Chaotic behavior, negative trends, and interpretation

Building upon our earlier findings regarding the system's dynamic responses to varying parameters, we now delve into the potential chaotic behavior and bifurcation analysis of our new electrical system. Notably, our exploration reveals significant insights, showcasing transitions to chaos for specific parameter values within the system. This investigation sheds light on the emergence of chaotic dynamics, complemented by negative trends observed in the system's behavior, culminating in a deeper interpretation of its intricate dynamics.

4.4.2 Identifying transitions to chaos

Our analysis has unveiled intriguing transitions to chaotic behavior within certain parameter ranges of the electrical system. Specifically, we would have identified distinct intervals—$0.5 < r < 0.6$, $0.7 < r < 0.8$, and $0.85 < r < 0.9$—where the system exhibits pronounced chaotic dynamics. This identification of chaos is substantiated by the observation of positive Lyapunov exponents, signifying sensitivity to initial conditions and confirming the system's transition to chaotic behavior [27].

4.4.3 Bifurcation analysis and negative trends

Moreover, our bifurcation analysis has prominently highlighted negative trends in the system's behavior. The majority of observed system behaviors manifest predominantly in the negative region of the x axis, indicating a clear trend toward negative outputs across explored r values. These negative curves hint at divergence toward negative magnitudes, suggesting decay, instability, or non-convergence within the system.

4.4.4 Interpreting chaos and trends

The convergence of positive Lyapunov exponents in the identified intervals, coupled with negative trends in the system's behavior, provides substantial evidence supporting the presence of chaotic dynamics. Understanding these intricate dynamics offers a pathway to deciphering the complexities within the electrical system, paving the way for potential control and application in diverse contexts.

4.4.5 Bifurcation plot with parameters

The bifurcation plot (**Figure 6**) illustrates the behavior of the electrical system with varying r values ($0.5 \le r \le 0.95$) and fixed parameters $R = 0.000025$, $L = 0.00045$, and $C = 0.73$. Notably, the observed behavior primarily concentrates in the negative region of the plotted values of x, suggesting potential instability or divergence within the system.

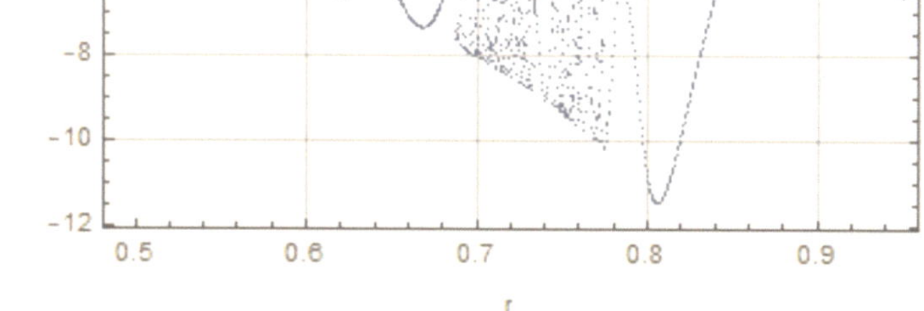

Figure 6.
Bifurcation plot showcasing system behavior with varying r values ($0.5 \le r \le 0.95$) and fixed parameters $R = 0.000025$, $L = 0.00045$, $C = 0.73$.

5. Methodologies

In this section, we elucidate the methodologies employed to assess the chaos and dynamical properties in the Yitang dynamics. Our approach combines numerical techniques with analytical methods to ensure a comprehensive understanding of the system's behavior.

5.1 Lyapunov exponent calculation

The Lyapunov exponents were calculated numerically using the computational capabilities of Mathematica software, which enabled precise evaluation of the system's sensitivity to initial conditions and insights into its chaotic behavior.

To quantify the system's sensitivity, we employed the following method to estimate the Lyapunov exponents for the Yitang dynamics (Eq. 1):

- *Initialization*: The initial conditions x_0, y_0, and z_0 were established in close proximity, allowing for the assessment of the divergence of nearby trajectories.

- *Numerical integration*: A numerical integration method, such as the Runge–Kutta method or Adams method, was utilized to solve the differential equations governing the Yitang dynamics.

- *Trajectory divergence*: At each iteration, the distance between two adjacent trajectories was computed, and the logarithm of this divergence was accumulated.

- *Lyapunov exponent calculation*: The Lyapunov exponent was estimated as the average of the logarithmic divergence over time, normalized by the iteration count and time step.

The computed Lyapunov exponents provided significant insights into the chaotic nature of the system, with non-zero values indicating chaotic dynamics.

6. Computation of Lyapunov exponents and entropy

The Lyapunov exponents (λ_i) are fundamental measures of sensitivity to initial conditions in dynamical systems. The following algorithm outlines the procedure used for computing these exponents:

Input: Dynamics function f, initial condition x_0, parameter α, number of iterations n, transient *tr*

Output: Lyapunov exponent

lyapunov(f, x_0, α, n, tr) $\leftarrow$;

begin

$df \leftarrow \text{Derivative}[1,0][f]$;

$\xi \leftarrow \text{NestList}[f[\#,\alpha]\&, x_0, n-1]$;

$\lambda \leftarrow \frac{1}{n}\sum \log|\text{df}[\#,\alpha]|\text{forDrop}[\xi, tr]$;

return λ;

end

Algorithm 1: Algorithm for Computing Lyapunov Exponents.

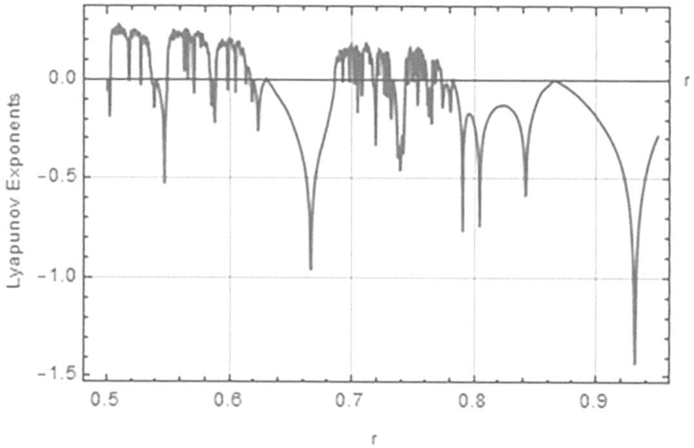

Figure 7.
Lyapunov exponents plot illustrating system behavior with varying r values $(0.5 \leq r \leq 0.95)$ *and fixed parameters* $R = 0.000025$, $L = 0.00045$, $C = 0.73$.

6.1 Lyapunov exponents plot with parameters

The Lyapunov exponents plot (**Figure 7**) reinforces our understanding by confirming transitions to chaos. The presence of positive Lyapunov exponents within specific intervals of r values validates the system's sensitivity to initial conditions and the emergence of chaotic behavior.

6.2 Implications and conclusion

The integration of bifurcation analysis, observations of negative trends, and confirmation through Lyapunov exponents plots solidifies our understanding of chaotic dynamics within the electrical system. This comprehensive analysis not only validates the presence of chaos but also presents opportunities for controlling and leveraging complex behaviors for various practical applications. The insights gained from this study could inform future research and engineering practices aimed at harnessing chaotic dynamics in innovative ways.

7. Prediction of limit cycles and fixed points

In this section, we analyze the phase portraits generated for the system dynamics, which provide crucial insights into the behavior of the electrical system under varying parameters. By plotting trajectories for different r values over 120 iterations, we can identify distinctive patterns that offer predictive cues regarding limit cycles and fixed points, enriching our understanding of the dynamical properties discussed in previous sections.

7.1 Observations from phase portraits

7.1.1 Periodicity in trajectories

The periodicity observed in the trajectories across different r values indicates the presence of limit cycles within the system. These trajectories, which recur in distinct patterns, signify the existence of cyclic behavior, suggesting that the system revisits certain states over specific intervals. The identification of these periodic regions is critical as they highlight stable, recurring states or limit cycles within the system's dynamics, aligning with our findings on chaotic behavior explored earlier.

7.1.2 Concentration in negative trend region

A notable concentration of trajectories in the negative trend region suggests the presence of fixed points within this domain. Fixed points typically denote states where the system stabilizes, remaining relatively unchanged over iterations. The accumulation of trajectories in this region indicates stable states or attractors toward which the system converges. Specifically, we identified an attracting fixed point at approximately $r = 0.7$ corresponding to $x = 1.54574$, indicating a robust stable state around this parameter.

7.1.3 Potential for chaotic regions

While the phase portraits predominantly highlight periodic and stable behavior, they also hint at the potential existence of chaotic regions. By extending the iterations to 150 or more, we may uncover irregular, seemingly random trajectories indicative of chaotic behavior. These regions could signify complex, non-repeating dynamics within the system, pointing to sensitivity to initial conditions and the presence of chaotic attractors that echo the chaotic properties identified through Lyapunov exponent analysis.

7.2 Predictive insights

Based on the observed patterns in the phase portraits, we can make predictions regarding limit cycles and fixed points. The periodicity in trajectories hints at the existence of stable limit cycles, while the concentration of trajectories in the negative trend region reinforces the presence of fixed points. Further exploration with increased iterations is likely to reveal chaotic regions, which would contribute significantly to a comprehensive understanding of the system's dynamics, building on the chaos analysis previously conducted.

7.3 Attracting fixed points and phase portrait

The system's attracting fixed points are summarized in **Table 1**. These values represent stable equilibrium points for various r values, highlighting the relationship between the parameter r and the system's behavior.

The phase portrait plot (**Figure 8**) visually represents the behavior of the system concerning these attracting fixed points. As illustrated, trajectories in the phase plane converge toward these stable states as r varies, confirming the attracting nature of these points.

In summary, the analysis of phase portraits reveals significant insights into the dynamics of the electrical system, particularly regarding limit cycles and fixed points. The observed periodicity and concentration of trajectories substantiate the presence of stable attractors. Furthermore, the potential identification of chaotic regions emphasizes the complexity of the system, warranting further investigation. These findings establish a foundational understanding of the system's dynamics, setting the stage for

0.5	1.44081
0.525	1.45602
0.55	1.47061
0.575	1.48459
0.6	1.49796
0.625	1.51074
0.65	1.52295
0.675	1.53461
0.7	1.54574
0.725	1.55635
0.75	1.56646
0.775	1.5761
0.8	1.58529
0.825	1.59404
0.85	1.60237
0.875	1.61031
0.9	1.61788
0.925	1.62508
0.95	1.63195

Table 1.
Subset of attracting fixed points for different r values.

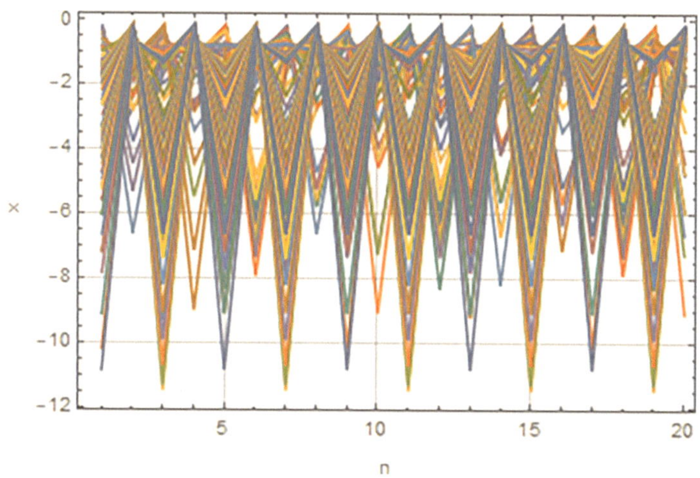

Figure 8.
Phase portrait showing trajectories converging to attracting fixed points. The plotted trajectories reflect the system's behavior, demonstrating convergence toward stable fixed points as r changes, which reinforces the predictions made regarding limit cycles and fixed points.

subsequent analyses, such as the frequency spectrum analysis discussed in the next section.

7.4 Fourier transform analysis

The Fourier transform of the electrical system's dynamics, characterized by parameters $r = 0.5$, $R = 0.000025$, $L = 0.00045$, and $C = 0.73$, is expressed as follows:

$$F(\omega) = 2.645898\delta(\omega)$$
$$+\frac{\sqrt{\frac{\pi}{2}}\left(C\,\mathrm{sgn}\,(\omega)\,{}_0F_1\left(;3;-\frac{2\pi\omega}{\mathrm{sgn}(\omega)}\right)(\omega\,\mathrm{sgn}(\omega)\,\mathrm{sgn}''(\omega)+2\,\mathrm{sgn}'(\omega)(\mathrm{sgn}(\omega)-\omega\,\mathrm{sgn}'(\omega)))\right.}{Cr^2\,\mathrm{sgn}\,(\omega)^4}$$
$$\left.+4\pi C_0\tilde{F}_1\left(;4;-\frac{2\pi\omega}{\mathrm{sgn}(\omega)}\right)(\mathrm{sgn}(\omega)-\omega\,\mathrm{sgn}'(\omega))^2+ir^2\,\mathrm{sgn}\,(\omega)^5\right).$$

This Fourier transform elucidates the frequency spectrum of the system's behavior, revealing the presence and distribution of frequencies within the trajectories. The Fourier transform is particularly significant in chaotic systems and electrical dynamics because it allows for the identification of dominant frequencies and harmonics, which are crucial for understanding the underlying mechanisms of chaotic behavior and energy dissipation in electrical circuits.

Figure 9 presents the plot of the Fourier transform derived from the system's phase portrait, showcasing the frequency distribution. Analyzing the Fourier transform provides insights into how various parameters influence the system's frequency-based behavior, essential for understanding stability and chaos in electrical systems. The presence of dominant frequency components may indicate resonance phenomena or chaotic oscillations, which are crucial for designing stable electrical systems and predicting their dynamic responses.

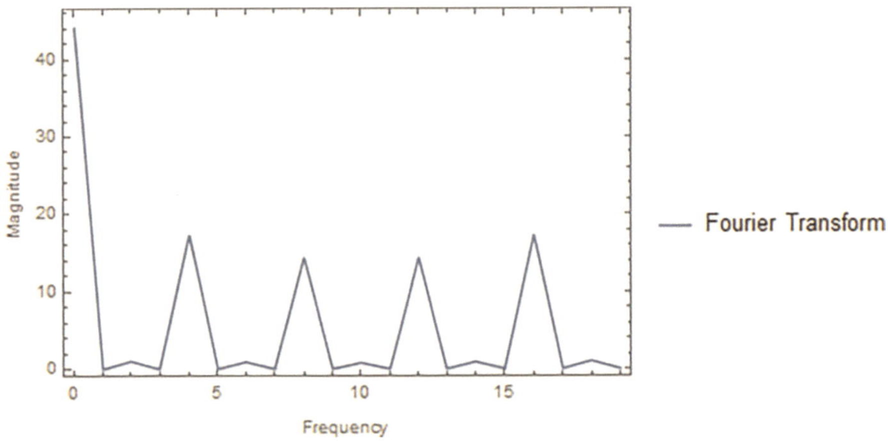

Figure 9.
Fourier transform plot of the system's dynamics.

7.5 Poincaré section analysis

The Poincaré section is a vital analytical tool in dynamical systems, enabling an efficient exploration of system behavior. It involves plotting the intersection points of trajectories with a selected surface in the phase space, providing a concise view of the system's dynamics without the need to analyze entire trajectories.

The Poincaré section plot illustrated in **Figure 10** offers insights into the system dynamics for specific parameters $R = 0.000025$, $L = 0.00045$, and $C = 0.73$. Each point on the plot corresponds to the system state at the intersections with the defined surface.

The plot visually represents the intersections of trajectories in phase space, illustrating recurrent states and visitation patterns on the chosen surface. The Poincaré section provides a condensed view of the system's dynamics, revealing insights into stability, periodicity, and potential chaotic regimes.

Notably, the observation of the Poincaré section predominantly in the negative region corresponds with earlier bifurcation patterns for the same parameter values, indicating critical implications for the system's dynamics. In electrical systems, negative regions might represent states or behaviors countering those observed in positive regions, suggesting the presence of stable states or attractors under specific parameter configurations.

The correlation between the negative Poincaré section and the bifurcation pattern implies a connection between the system's stability and behavior across different parameter regimes. This finding indicates that the system may exhibit contrasting or complementary behaviors in these regions, signifying diverse operational states.

Further exploration of the dynamics within the negative Poincaré section, in conjunction with bifurcation analyses, could provide deeper insights into the system's response to varying parameter configurations.

7.6 Poincaré sections and phase analysis for varied parameter values

In this subsection, we include Poincaré section plots and phase analysis for various parameter configurations. For each case, we adjust parameters R, L, and C while keeping r within the corresponding range.

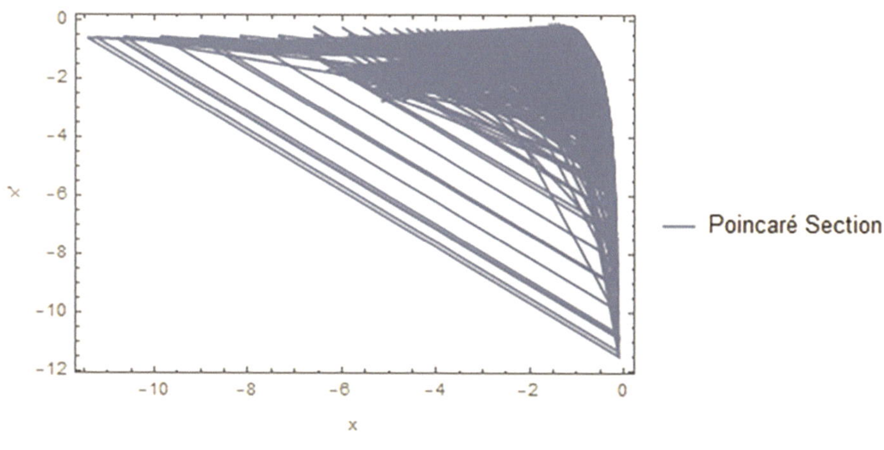

Figure 10.
Poincaré section for the electrical system with $R = 0.000025$, $L = 0.00045$, and $C = 0.73$.

Case 1: $R = 0.000025$, $L = 0.00045$, $C = 1.5$ (*r* **range: 0.5 to 0.95**)

The Poincaré section plotted for the selected parameters provides insight into the system's behavior, capturing intersection points in phase space that indicate recurrent states or areas of convergence (**Figure 11**).

The presence of intersections in specific regions of the phase space suggests recurrent behaviors or attractors within the system. Additionally, the blue intersections in the positive region indicate the potential for chaotic behavior, characterized by sensitivity to initial conditions and non-repeating dynamics.

Case 2: $R = 0.000025$, $L = 0.00045$, $C = 4.5$ (*r* **range: 0.5 to 0.95**)

In the second case, with adjusted parameters, the Poincaré section analysis reveals intriguing observations (**Figure 12**).

All plotted intersections appear in the positive region, indicative of distinct behavior compared to previous cases. The absence of intersections in the negative region, coupled with blue intersections solely in the positive region, suggests a potential for chaotic behavior.

This aligns with chaotic dynamics characterized by sensitivity to initial conditions and non-repeating trajectories. The absence of intersections in the negative region

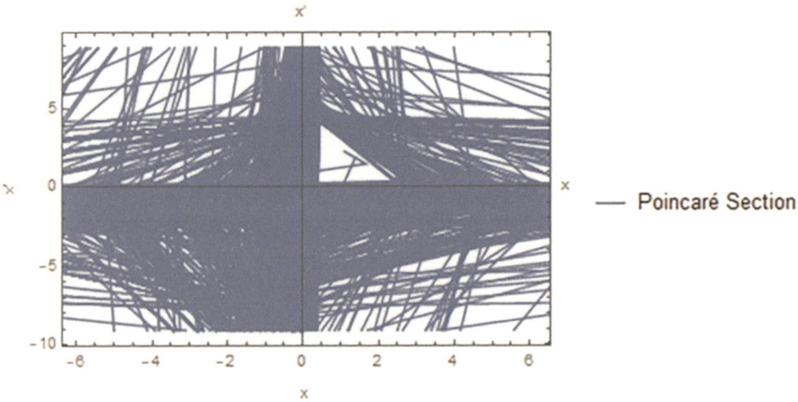

Figure 11.
Poincaré section for R = 0.000025, L = 0.00045, C = 1.5 (r range: 0.5 to 0.95).

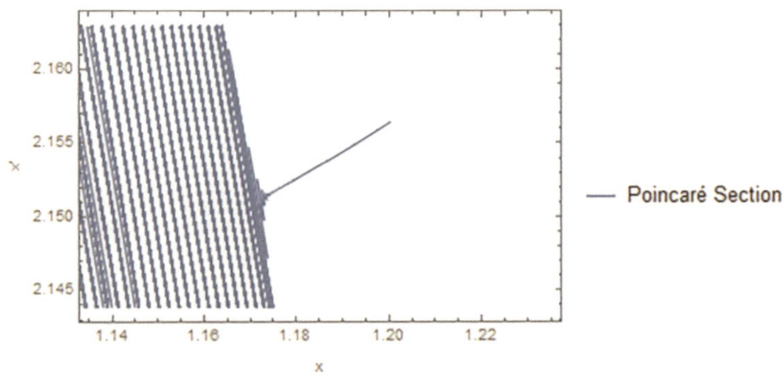

Figure 12.
Poincaré section for R = 0.000025, L = 0.00045, C = 4.5 (r range: 0.5 to 0.95).

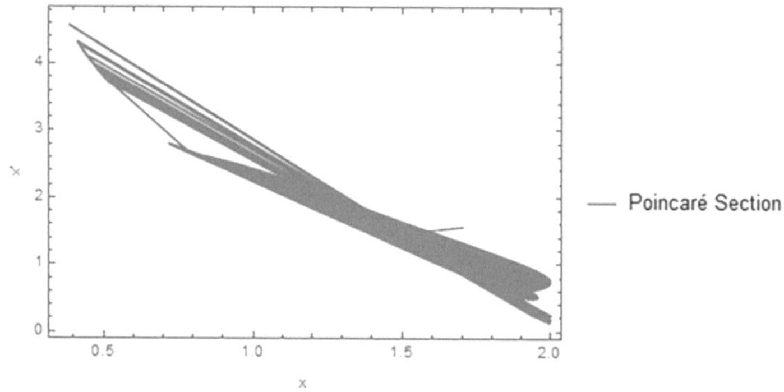

Figure 13.
Poincaré section for $R = 800$, $L = 800$, $C = 800$ (r range: 0.5 to 0.95).

may indicate a lack of convergence toward stable states, underscoring the system's propensity for irregular behavior.

Case 3: $R = 800$, $L = 800$, $C = 800$ (*r* **range: 0.5 to 0.95**)

For the third case, with significantly increased parameter values, the Poincaré section illustrates compelling insights (**Figure 13**).

All intersections occurring in the high positive region align with chaotic behavior patterns. The presence of intersections exclusively in this region signifies erratic and non-repeating dynamics, typical of chaotic systems.

The high parameter values drive the system into a state of increased complexity and sensitivity to initial conditions, resulting in intricate trajectories. The absence of intersections in the negative region reinforces the chaotic nature, depicting the lack of convergence toward stable states.

In conclusion, the combined analysis of Fourier transforms and Poincaré sections provides critical insights into the chaotic dynamics and stability of electrical systems. The ability to visualize how parameters influence behavior enhances our understanding of system dynamics, paving the way for potential applications in control systems and circuit design.

This observation further reinforces the chaotic tendencies identified in this electrical system, underscoring its sensitivity to variations in parameters and initial conditions. The Fourier transform analysis not only reveals the frequency spectrum of the system's dynamics but also indicates how these frequencies respond to changes in the system's parameters, which can amplify or mitigate chaotic behaviors. The presence of dominant frequencies often correlates with specific parameter regimes that may lead to chaotic oscillations or stable behavior, illustrating the intricate interplay between the system's dynamics and its response to perturbations.

For a comprehensive exploration of the diverse behaviors within our electrical system model, readers are encouraged to consult the accompanying Mathematica code available in our notebook. The sections titled "*Bifurcation Analysis*" and "*Lyapunov Exponents*" include code snippets and visualizations for bifurcation diagrams and Lyapunov exponent calculations, providing valuable insights into the stability and chaotic characteristics of the system. Additionally, the section on "*Poincaré Section Plots*" offers further analysis of the system's dynamics, illustrating how trajectories evolve in phase space and revealing the nature of recurrent states.

8. Strange attractor and transition to voltage collapse

In this section, we explore the strange attractor of our newly proposed dynamical system:

$$x_{n+1} = 1 - \left(\frac{\sin(\pi/x_n)}{r\pi/x_n}\right)^2 + \frac{R}{L} + \frac{1}{Cx_n},$$

where x_n represents the state of the system, and R, C, L, and r are the electrical parameters (resistance, capacitance, inductance, and a scaling factor). Using the parameter values $R = 1.0$, $L = 1.5$, $C = 0.5$, and $r = 0.8$, we numerically simulate the system over 10,000 iterations.

The 3D visualization of the strange attractor, shown in **Figure 14**, reveals the intricate geometry of the system's chaotic state. The x_n-axis represents the system state, while the auxiliary transformations $y = \sin(x_n)$ and $z = \cos(x_n)$ highlight the attractor's structure. The plot demonstrates sensitivity to initial conditions, characteristic of chaotic systems, and confirms the presence of a strange attractor under these parameter values.

This analysis provides critical insights into the system's stability. As parameters such as R/L and $1/(Cx_n)$ increase, the strange attractor begins to exhibit signs of instability, leading to a transition toward *voltage collapse*. Voltage collapse occurs when the balance between the system's reactive and damping components is disrupted, resulting in a divergence of the trajectories from the attractor. In physical terms, this

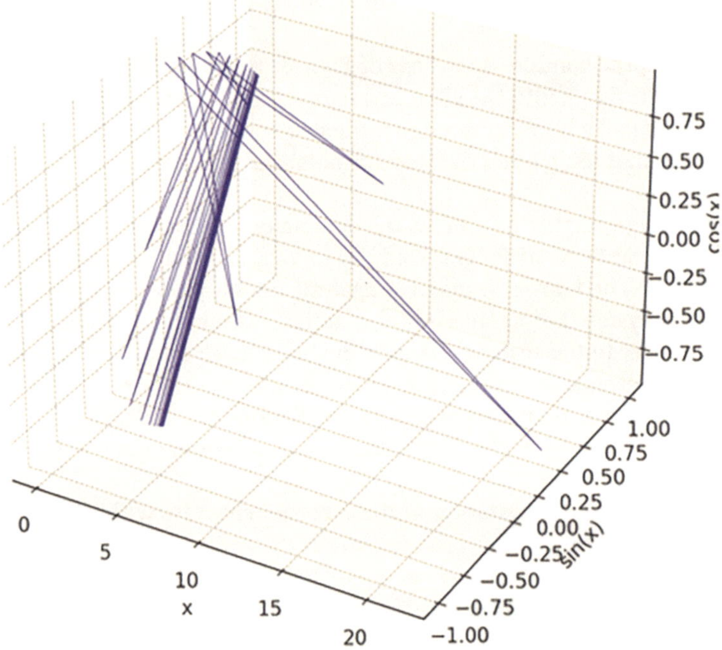

Figure 14.
3D strange attractor of the dynamical system. The attractor is plotted using $R = 1.0$, $L = 1.5$, $C = 0.5$, $r = 0.8$, *and an initial condition* $x_0 = 0.1$.

reflects an inability of the system to maintain equilibrium, ultimately causing the electrical network to fail.

The observations from the strange attractor's behavior indicate that the onset of voltage collapse can be predicted by monitoring changes in the attractor's geometry, particularly its sensitivity to small parameter variations. Such results establish a compelling link between the chaotic dynamics of the system and practical phenomena like voltage collapse in electrical circuits.

9. Energy distribution in our new electrical system

For our electrical system, described by the equation

$$f(x, r, R, L, C) = 1 - \left(\frac{\sin\left(\frac{\pi}{x}\right)}{r\pi/x}\right)^2 + \frac{R}{L} + \frac{1}{Cx},$$

the following parameter values were employed:

$$R = 0.000025,$$
$$L = 0.00045,$$
$$C = 0.73.$$

This system, incorporating resistors, inductors, and capacitors, is significantly influenced by the parameter r, which affects the energy storage behavior throughout the circuit.

9.1 Interpretation of r

The parameter r plays a crucial role in shaping the system's dynamics, particularly concerning energy storage. Variations in r lead to changes in the electrical characteristics, including the stability of the system and its resonant behaviors. As r increases, the energy distribution among the components can shift, emphasizing the interplay between system parameters and energy dynamics.

9.2 Behavior of energy vs. r

To analyze the energy behavior in relation to r, we plot the energy stored in the inductor across a spectrum of values (**Figure 15**):

The plot illustrates the variation of energy stored in the inductor with respect to r. It highlights the sensitivity of energy distribution to changes in r, revealing how the system adapts to different parameter settings.

This variation in energy distribution provides essential insights into the behavior of the system, aiding our understanding of how r impacts the electrical dynamics and energy storage mechanisms.

9.3 Energy distribution with varying capacitance

In our electrical system, we set parameters $R = 2$, $L = 6$, and varied C within the range $(0.000000073, 0.0000073]$. We examine the relationship between the capacitance C and the energy stored in the inductor.

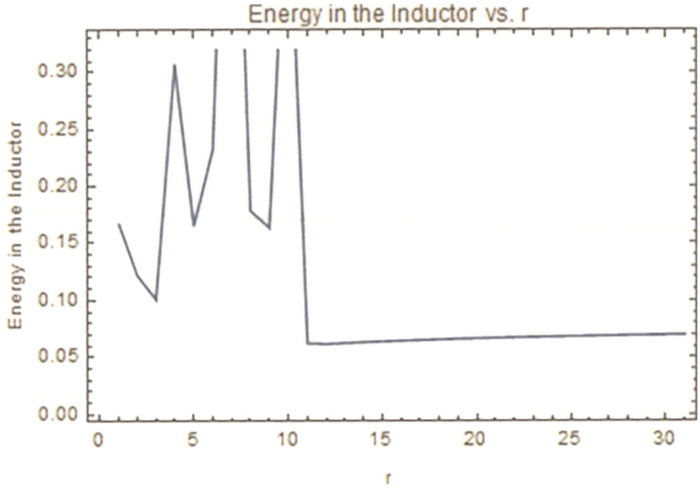

Figure 15.
Energy in the inductor vs. r for R = 0.000025, *L* = 0.00045, *and C* = 0.73.

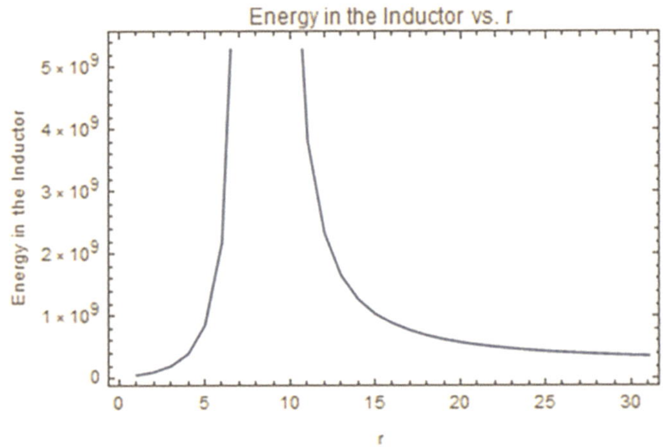

Figure 16.
Energy in the inductor vs. r for varying C values.

9.3.1 Effect of C *on energy in the inductor*

The following plot demonstrates the correlation between the stored energy in the inductor and parameter r while varying C (**Figure 16**):

As C decreases, a significant increase in the energy stored in the inductor is observed. This behavior reflects a vital interaction between capacitance and the energy dynamics of the electrical system. The sharp discontinuity at very small C values indicates a critical threshold where the system's behavior experiences a notable change.

10. Energy transfer analysis

Energy transfer within electrical systems refers to the distribution and movement of energy among components such as capacitors, inductors, and resistors. Analyzing

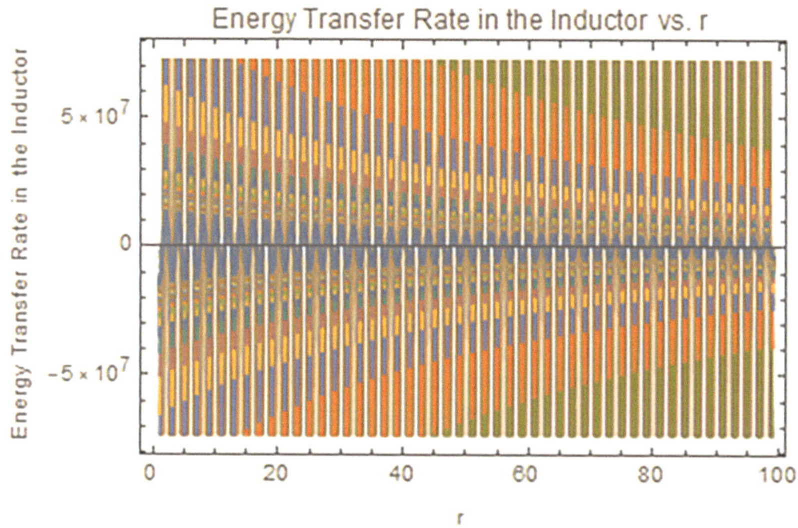

Figure 17.
Energy transfer rate in the inductor vs. r (Parameters: $R = 2.000025$, $L = 6.00045$, $C = 0.000000073$).

how energy is allocated and its changes with varying parameters is fundamental for understanding the system's dynamics.

In our investigation, we explored energy transfer rates within the system, focusing on energy stored in the components over time. The plot in **Figure 17** illustrates the rate of energy transfer to the inductor concerning the parameter *r*, varying between 0.5 and 2.

This plot reveals the variability in the energy transfer rate to the inductor for different values of *r*. It underscores the relationship between parameter variations and energy flow dynamics, contributing to our understanding of energy distribution within the system.

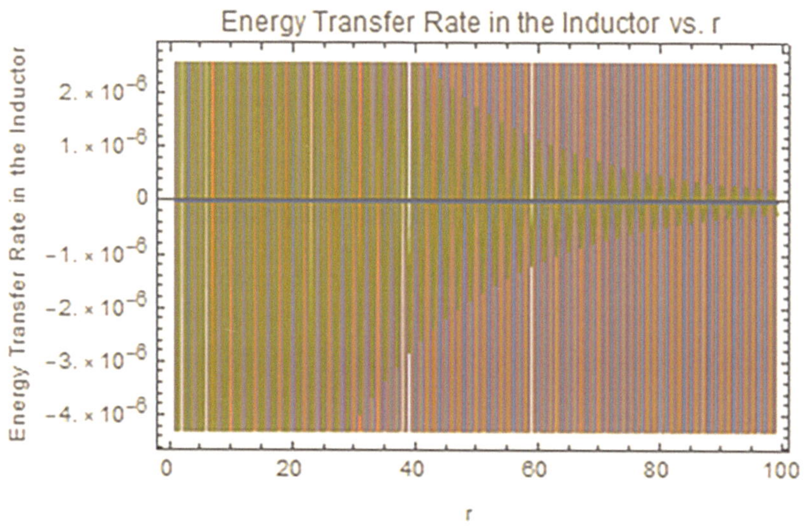

Figure 18.
Energy transfer rate in the inductor vs. r for $C = 0.73$ (Parameters: $R = 2.000025$, $L = 6.00045$, $C = 0.73$).

Further investigation into the influence of capacitance (C) on energy transfer was also conducted. The plot in **Figure 18** illustrates how increasing C from 0.000000073 to 0.73 affects the energy transfer rate in the inductor [20].

As shown in **Figure 18**, increasing the capacitance (C) leads to a decrease in the energy transfer rate. This change induces both positive and negative values in the energy transfer rate, suggesting a reconfiguration of energy flow and distribution within the system.

11. Chaotic behavior, energy transfer, and storage in electrical systems

In electrical systems, interactions between system parameters and nonlinear components can often result in chaotic behavior. Although no specific theorem directly links chaos with energy transfer and storage, certain behaviors can be correlated, particularly in systems exhibiting intricate dynamics.

The derived electrical system formula: The system is defined by the equation:

$$f(x, r, R, L, C) = 1 - \left(\frac{\sin\left(\frac{\pi}{x}\right)}{r\frac{\pi}{x}}\right)^2 + \frac{R}{L} + \frac{1}{Cx}.$$

Connecting Chaos and Energy in Electrical Systems:

• *Parameter sensitivity*: The system's dynamics exhibit sensitivity to variations in parameters, such as capacitance (C). Changes in C may lead to bifurcations and abrupt behavioral changes, potentially uncovering chaotic regions.

• *Energy dynamics and chaotic behavior*: Nonlinear components in the system contribute to unpredictable behavior. Reducing C values may induce chaotic responses, resulting in irregular energy distributions and erratic energy transfer rates.

• *System complexity*: Chaotic behavior often arises from the complexity of the system. As C changes, the system may enter chaotic regions characterized by unpredictable energy distributions and intricate energy transfer dynamics.

While a direct theorem linking chaos to energy transfer/storage in electrical systems is lacking, the observed behaviors align with principles of chaos theory. Sensitivity to parameter changes, nonlinear dynamics, and system complexity contribute to the chaotic behavior, impacting energy transfer and storage patterns.

12. Comparison of Yitang and Montgomery dynamics

This section provides a comparative analysis of the quantum properties of Yitang and Montgomery dynamics through their respective Hamiltonian operators and eigenvalues.

12.1 Yitang dynamics

The quantum Hamiltonian operator for Yitang dynamics is expressed as

$$\hat{H}_{\text{Yitang}} = -\frac{\hbar^2}{2m}\nabla^2 + V(z),$$

where ∇^2 represents the Laplacian operator, and $V(z)$ denotes the potential energy associated with the Yitang function.

12.1.1 Hamiltonian operator

The Yitang Hamiltonian operator is formulated as

$$\hat{H}_{\text{Yitang}} = -\frac{1}{2m}\frac{d^2}{dz^2} - \frac{c^2}{mz^{2\alpha}}(\log(z))^{-2\alpha},$$

with potential energy terms given by

$$\psi_1(z) = \frac{d}{dz}\log(z) \quad \text{and} \quad \psi_2(z) = (\log(z))^{-\alpha}.$$

12.1.2 Eigenvalues

The eigenvalues for the Yitang dynamics were calculated using the Implicitly Restarted Arnoldi Method (IRAM). The results indicate complex eigenvalues that exhibit sensitivity to parameter variations.

12.2 Montgomery dynamics

The Montgomery dynamics is characterized by the Hamiltonian operator

$$\hat{H}_{\text{Montgomery}} = -\frac{\hbar^2}{2m}\nabla^2 + V(x),$$

where $V(x)$ is the potential energy corresponding to the Montgomery function.

12.2.1 Hamiltonian operator

The Montgomery Hamiltonian operator is defined as

$$\hat{H}_{\text{Montgomery}} = -\frac{1}{2m}\frac{d^2}{dx^2} + V(x),$$

which includes the potential energy term associated with the Montgomery dynamics.

12.2.2 Eigenvalues

Similar to Yitang dynamics, the eigenvalues of Montgomery dynamics were computed using the IRAM. Notably, the obtained eigenvalues are real, suggesting a distinct quantum behavior compared to the Yitang dynamics.

12.3 Implicitly restarted Arnoldi method (IRAM)

The Implicitly Restarted Arnoldi Method (IRAM) is an iterative algorithm designed for computing a limited number of eigenvalues and corresponding

Dynamics	Eigenvalue Index	Eigenvalues
Yitang	1	$7.11343 - 0.178929i$
Montgomery	1	-2.99007

Table 2.
Comparison of eigenvalues for Yitang and Montgomery dynamics.

eigenvectors of large, sparse matrices. It is particularly effective for quantum systems, where direct diagonalization can be computationally prohibitive. By iteratively refining eigenvalue approximations, IRAM serves as a powerful tool for exploring quantum dynamics.

12.4 Comparison of eigenvalues

Table 2 summarizes the computed eigenvalues for both Yitang and Montgomery dynamics.

12.5 Diagonalizability

The diagonalizability of Hamiltonian operators is crucial for solving quantum systems. A comparative analysis of the diagonalizability of $\hat{H}_{\text{Yitang}}$ and $\hat{H}_{\text{Montgomery}}$ follows:

12.5.1 Yitang dynamics

The complex eigenvalues associated with Yitang dynamics indicate the potential for complex eigenvectors. Further investigation is required to assess the diagonalizability of $\hat{H}_{\text{Yitang}}$.

12.5.2 Montgomery dynamics

In contrast, the real eigenvalues of Montgomery dynamics suggest that the corresponding eigenvectors are likely real and linearly independent, making $\hat{H}_{\text{Montgomery}}$ expected to be diagonalizable.

In conclusion, while Yitang dynamics reveals complex eigenvalues, Montgomery dynamics is more likely to exhibit diagonalizability due to its real eigenvalues.

13. Schrödinger equation with approximated potential

The Hamiltonian operator is a fundamental construct in quantum mechanics, representing the total energy of a system. It plays a pivotal role in governing the dynamics and behavior of quantum systems, particularly through the Schrödinger equation, which describes how the quantum state of a physical system evolves over time.

The Schrödinger equation, defined with a specific potential function, allows for the derivation of eigenvalues and eigenfunctions, which are essential for understanding energy levels and the probabilistic nature of quantum particles. These eigenfunctions form a complete basis set for the Hilbert space, enabling the description of any quantum state as a superposition of these basis states.

Beyond its significance in quantum mechanics, the Hamiltonian framework has important applications in signal processing. The eigenvalues and eigenfunctions derived from the Hamiltonian can relate to frequency components and spectral analysis in signal processing, thereby facilitating the analysis of complex signals and systems through techniques derived from quantum mechanics, such as Fourier analysis and wavelet transforms.

Furthermore, the concept of the Hamiltonian extends into electrical systems within electrical engineering. It can be employed to model and analyze the optimality and robustness of electrical circuits and systems. By characterizing energy storage and dissipation within circuits, the Hamiltonian contributes to optimizing circuit performance and ensuring system stability under various conditions.

14. Solving the Schrödinger equation

In this section, we solve the Schrödinger equation using the approximated potential function:

$$V(x) = A + \frac{1}{Cx}, \tag{4}$$

where

$$A = 1 - \frac{1}{r^2} + \frac{R}{L}. \tag{5}$$

The time-independent Schrödinger equation is given by:

$$-\frac{\hbar^2}{2m}\frac{d^2\psi(x)}{dx^2} + V(x)\psi(x) = E\psi(x). \tag{6}$$

Substituting the approximated potential $V(x)$, we obtain:

$$-\frac{\hbar^2}{2m}\frac{d^2\psi(x)}{dx^2} + \left(A + \frac{1}{Cx}\right)\psi(x) = E\psi(x). \tag{7}$$

Rearranging this, we get:

$$\frac{d^2\psi(x)}{dx^2} = -\frac{2m}{\hbar^2}\left(E - A - \frac{1}{Cx}\right)\psi(x). \tag{8}$$

Define a new variable:

$$k^2(x) = \frac{2m}{\hbar^2}\left(E - A - \frac{1}{Cx}\right). \tag{9}$$

Thus, the Schrödinger equation simplifies to:

$$\frac{d^2\psi(x)}{dx^2} = -k^2(x)\psi(x). \tag{10}$$

In the special case where $k^2(x)$ is approximately constant, the solution to the differential equation is given by:

$$\psi(x) = C_1 \sin(kx) + C_2 \cos(kx), \tag{11}$$

where $k = \sqrt{\frac{2m}{\hbar^2}(E - A)}$.

For the general case where $k^2(x)$ varies with x, the exact solution typically involves special functions such as the Whittaker functions or hypergeometric functions. These functions can be obtained numerically or through analytical techniques in mathematical physics.

The next step involves using numerical methods or special functions to find the exact eigenfunctions and eigenvalues for specific parameters in practical scenarios.

15. Numerical simulation and analysis

In this section, we analyze the behavior of the potential function $V(x)$ and the corresponding eigenfunction $\psi(x)$ for a quantum system governed by the Schrödinger equation. The potential function is defined as:

$$V(x) = A + \frac{1}{Cx},$$

where A and C are constants specific to the system. The eigenfunction, obtained as an analytical solution to the Schrödinger equation, is given by:

$$\psi(x) = C_1 \sin(kx) + C_2 \cos(kx),$$

where k is the wavenumber related to the energy level E and the mass m of the particle.

15.1 Motivation

Understanding the behavior of quantum systems through their potential functions and eigenfunctions is critical in both quantum mechanics and signal processing. The potential function $V(x)$ determines the dynamics of the particle within the quantum well, while the eigenfunction $\psi(x)$ provides insight into the probability distribution of the particle's position.

In signal processing, these eigenfunctions can be interpreted as signals, where their frequency components are directly tied to the physical parameters of the system. Analyzing these signals can reveal key properties of the system, such as stability, robustness, and optimality. Specifically, in the context of electrical systems, such analysis can guide the design of circuits that minimize energy loss and maximize efficiency, ensuring the system's optimal performance under various operating conditions.

Parameter	Value
A	0.06
C	0.73
m	3.8 (arbitrary units)
$\hbar$	1 (arbitrary units)
E	0.5 (example value)
C_1	1
C_2	1.7

Table 3.
Parameters used for the numerical simulation.

15.1.1 Numerical simulation

To illustrate this, we plot the potential function $V(x)$ and the corresponding eigenfunction $\psi(x)$ using the parameters listed in **Table 3**. The eigenfunction exhibits a sinusoidal pattern, indicating a bound state in the potential well. The amplitude and wavelength of the oscillations depend on the parameters A, C, m, and E, as shown in **Figure 19**.

The sinusoidal nature of the eigenfunction suggests that the particle is likely confined within a potential well, with oscillations indicating discrete energy levels. Further analysis, such as Fourier transform or wavelet analysis, can provide deeper insights into the properties of the eigenfunction, allowing for the decomposition of the signal into its constituent frequency components. This is particularly useful in signal processing applications, where identifying and isolating specific frequencies can enhance system performance and robustness.

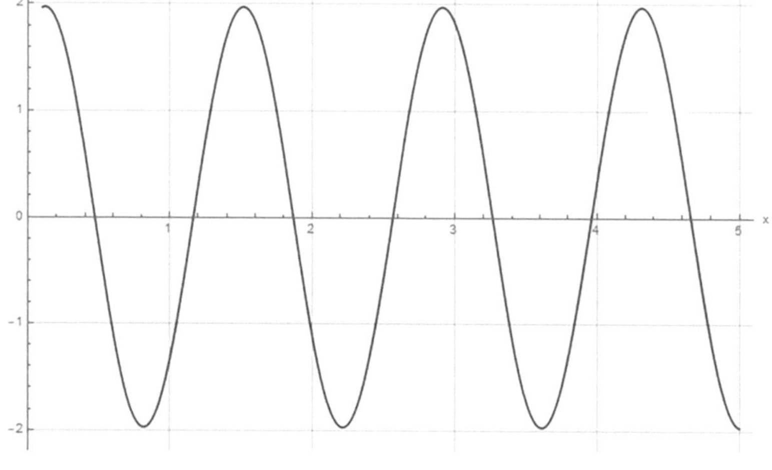

Figure 19.
Plot of the potential function and eigenfunction.

16. Conclusion

In this comprehensive investigation of the electrical system's dynamics, we illuminated various facets of its behavior under diverse conditions. The system, characterized by a complex equation involving critical parameters such as r, R, L, and C, exhibited intricate patterns and behaviors, providing valuable insights into its stability and dynamical properties.

The phase portraits revealed periodic trajectories indicative of stable limit cycles, alongside regions of convergence toward attracting fixed points, suggesting equilibrium states that are pivotal for system stability. Notably, the hints of chaotic behavior encourage further exploration into the dynamics of the system, particularly within engineering applications where chaos can significantly influence performance and reliability [28].

Exploring attracting fixed points elucidated stable equilibrium states for various r values, affirming the system's behavior in response to parameter variations. The trajectories converging toward these points not only reinforce the notion of stability but also emphasize the importance of understanding these dynamics for effective system design.

Our frequency spectrum analysis, employing the Fourier transform, revealed crucial insights into the distribution of frequencies within the trajectories, highlighting dominant frequencies and their role in the system's overall dynamics. This understanding is essential in engineering contexts where frequency response plays a vital role in the design and optimization of electrical systems.

The Poincaré section analysis effectively condensed the system's behavior, showcasing recurring states that aid in understanding stability, periodicity, and potential chaos. The correlation between the Poincaré section and bifurcation patterns provides critical insights into the diverse operational states that can emerge within the system, which are crucial for engineers aiming to harness or mitigate chaotic behavior.

In examining energy distribution, we emphasized the sensitivity of energy storage to parameters like r and C. Our findings demonstrated that changes in r led to variations in energy levels, while a decrease in C significantly increased inductor energy, thereby underscoring the intricate relationship between capacitance and energy dynamics. This insight is vital for engineers designing systems that require precise energy management.

Furthermore, our analysis of energy transfer illuminated how alterations in r and C impact the rate of energy transfer to the inductor, providing foundational knowledge essential for optimizing energy flow and distribution in practical applications.

In the context of quantum dynamics, the comparison between Yitang and Montgomery dynamics uncovered distinctive characteristics, with the latter exhibiting real eigenvalues that enhance its appeal for tackling the Riemann hypothesis [29]. The alignment of the system's quantum behavior with diagonalizability requirements offers intriguing prospects for future research at the intersection of classical dynamics and number theory [30].

Looking ahead, our work opens several avenues for future research. Investigating the connections between chaotic dynamics and practical applications in electrical engineering could lead to innovative designs that leverage chaotic behavior for enhanced system performance. Additionally, integrating our findings with recent advancements in nonlinear dynamics and control theory may yield new strategies for managing instability in complex systems. Furthermore, exploring the implications of our work on the Riemann hypothesis invites a deeper investigation into the interplay between number theory and quantum mechanics, potentially leading to groundbreaking insights in both fields.

In conclusion, this multifaceted analysis not only deepens our understanding of the studied electrical system but also underscores the critical importance of dynamical systems in engineering applications. The exploration of chaotic behaviors within electrical systems holds promise for advancing system design and optimization. Our results highlight the significance of deriving such dynamical systems, particularly as they translate into electrical systems, thus paving the way for innovations in both theoretical and applied fields [31–36].

17. Future research

The exploration of chaotic dynamics in electrical systems offers a promising avenue for future research, particularly in the context of circuit applications. As highlighted by Lai et al. in their work on the circuit application of chaotic systems, there is a growing interest in modeling, analyzing, and controlling chaotic behaviors within circuit frameworks [37]. This presents a fertile ground for developing innovative chaotic systems that can enhance the functionality and performance of electrical circuits.

Future research could focus on the derivation of new chaotic systems specifically tailored for circuit applications. This involves not only extending the current understanding of chaotic behavior in existing electrical systems but also applying novel mathematical frameworks and computational techniques to design and analyze new chaotic circuits. The integration of chaotic systems into electrical circuits has the potential to improve their robustness, efficiency, and response characteristics, particularly in environments where traditional systems may falter.

Key areas of investigation may include:

- *Modeling new chaotic systems*: Developing mathematical models for new chaotic systems that can be implemented in circuit designs. This could involve exploring alternative nonlinear components or configurations that yield richer dynamics and more complex behaviors.

- *Dynamical analysis*: Conducting comprehensive dynamical analyses of these newly derived chaotic systems to understand their stability, bifurcations, and transitions between different dynamical regimes. This analysis will be crucial in identifying operational parameters that can harness chaos for practical applications.

- *Control strategies*: Investigating advanced control strategies for managing chaotic behaviors in circuits. Effective control methods could mitigate undesirable chaotic dynamics while retaining beneficial characteristics, leading to enhanced circuit performance.

- *Experimental validation*: Collaborating with experimental physicists and engineers to validate theoretical predictions and models through practical implementations. Building prototype circuits that embody these new chaotic systems would provide invaluable insights into their real-world behavior.

- *Applications in engineering*: Exploring the implications of these chaotic circuits in various engineering applications, such as secure communication systems, signal processing, and sensor technologies. Harnessing the unique properties of chaos could lead to breakthroughs in the efficiency and effectiveness of these applications.

- *Connection to number theory*: The derivation of chaotic systems from conjectures in number theory represents a novel and efficient method for analyzing complex behaviors in electrical circuits. By leveraging discrete chaotic maps, researchers can uncover insights that were previously difficult to obtain. This approach not only enhances our understanding of chaos in dynamical systems but also opens up avenues for practical applications.

- *Practical applications of entropic signals*: Further investigation into the practical application of entropic signals, as discussed in the context of chaotic electrical systems, can provide new methodologies that utilize the unique properties of these signals. Recent works, such as the comprehensive review by Petrzela [4], emphasize the significance of chaos in analog circuits and present potential practical applications, which could greatly benefit our understanding and utilization of chaotic dynamics.

By aligning future research efforts with the latest advancements in chaotic systems, as emphasized by Lai et al., we can drive innovations that not only deepen our understanding of dynamical systems but also open new frontiers in electrical engineering. As the field of chaotic circuits continues to evolve, integrating theoretical, computational, and experimental approaches will be vital in realizing the potential of chaos in practical applications, ultimately contributing to the advancement of technology and engineering solutions.

Acknowledgements

I would like to express my sincere gratitude to the anonymous referee whose insightful comments and constructive feedback greatly contributed to the refinement of this work. Your dedication to maintaining the quality and integrity of the content has been invaluable, and I sincerely appreciate the time and effort you invested in reviewing my work.

I am profoundly thankful to my family for their unwavering support throughout this journey. My heartfelt appreciation goes to my parents for instilling in me a love for knowledge and a strong work ethic. To my loving wife, your encouragement and understanding have been my pillars of strength, and I am grateful for the sacrifices you have made to see me succeed.

I extend my love and gratitude to my sons, Taha Abdeljalil and Taki Abdessalem, for being a source of inspiration and joy in my life. Your presence fuels my determination to strive for excellence, and I am blessed to have you by my side.

A special acknowledgment is due to my dear friend, Bouzari Abdelkader, a dedicated teacher of physics in high school. Your support, encouragement, and intellectual insights have played a significant role in shaping my ideas and refining my work. Your friendship has been a constant source of motivation, and I am grateful for the positive influence you have had on my academic and personal growth.

I would also like to extend my gratitude to my dear sister Chaima (Arrass Ta3i) and my sister Widad, as well as my brothers Saddam and Kheireddine, Foued (Jouad), for their unwavering support and encouragement.

Data availability

The data used in this study is publicly available and sourced from the following paper, which serves as the basis for our research [27].

All data and methodologies used in our analysis are in accordance with the methods outlined in the aforementioned paper. Additionally, any supplementary data required for replication or further analysis is available upon request from the corresponding author.

Declarations

The authors declare that there is no conflict of interest regarding the publication of this chapter. We confirm that this research was conducted in an unbiased and impartial manner, without any financial, personal, or professional relationships that could be perceived as conflicting with the objectivity and integrity of the research or the publication process.

A. Derivation of Chaotic operator for Montgomery dynamics

The Montgomery Dynamics is described by the equation:

$$x_{n+1} = \frac{1}{2}x_n + \frac{1}{4x_n} - \frac{\alpha}{x_n^3}, \quad x_0 \neq 0 \tag{12}$$

A.1 Hamiltonian operator

The quantum Hamiltonian operator for the Montgomery dynamics is given by:

$$\hat{H} = -\frac{\hbar^2}{2m}\nabla^2 + V(x) \tag{13}$$

where ∇^2 is the Laplacian operator, and $V(x)$ represents the potential energy associated with the Montgomery function.

A.2 Potential energy terms

Express the potential energy term as:

$$V(x) = -\frac{\alpha^2}{2mx^2} - \frac{1}{4m}\psi_1(x) + \frac{\alpha}{mx^3}\psi_2(x) \tag{14}$$

where $\psi_1(x) = \frac{\partial}{\partial x}\left(\frac{1}{x}\right)$ and $\psi_2(x) = \frac{1}{x^3}$.

A.3 Chaotic operator

The derived chaotic operator is given by:

$$\hat{C}_{\text{chaotic}} = \hat{A} \cdot \nabla^2 + \hat{B} \cdot \frac{1}{4m}\psi_1(x) + \hat{D} \cdot \frac{\alpha}{mx^3}\psi_2(x) \tag{15}$$

B. Extension to Riemann hypothesis

B.1 Logarithmic term and Von Mangoldt function

Incorporate the Von Mangoldt function into the chaotic operator:

$$\hat{C}_{\text{chaotic}} = \hat{A} \cdot \nabla^2 + \hat{B} \cdot \frac{1}{4m}\psi_1(x) + \hat{D} \cdot \frac{\alpha}{mx^3}\psi_2(x) + \hat{E} \cdot \Lambda(x) \tag{16}$$

where $\Lambda(x)$ is the Von Mangoldt function.

B.2 Rationale

Discuss the rationale behind incorporating the Von Mangoldt function and its connection to addressing the Riemann Hypothesis.

B.3 Numerical simulations

Perform numerical simulations to analyze the properties and behavior of the modified chaotic operator.

B.4 Comparison with Riemann Zeta function

Compare the results obtained from the modified chaotic operator with the expected behavior based on the Riemann zeta function.

B.5 Insights and applications

Discuss insights gained into quantum chaos and the suitability of Montgomery dynamics in addressing the Riemann Hypothesis.

DOI: http://dx.doi.org/10.5772/intechopen.1010620

Author details

Zeraoulia Rafik[1*], Alvaro Humberto Salas[2] and Ayadi Souad[3]

1 Acoustics and Civil Engineering Laboratory, Faculty of Material Sciences and Computer Science, Mathematics Department, Khemis Miliana University, Khemis Miliana, Algeria

2 Departamento de Matemáticas, Universidad Nacional de Colombia, Bogotá, Colombia

3 Acoustics and Civil Engineering Laboratory, Faculty of Material Sciences and Computer Science, Physics Department, Khemis Miliana University, Khemis Miliana, Algeria

*Address all correspondence to: zeraoulia@univ-dbkm.dz

References

[1] Banerjee S, Rondoni L. Applications of Chaos and Nonlinear Dynamics in Science and Engineering - Vol. 4. Understanding Complex Systems. Berlin, Germany: Springer; 2015. DOI: 10.1007/978–3–319–17037–4

[2] Rafik Z. Derivation of chaotic dynamics from Montgomery conjecture and its interpretation as an electrical system. arXiv:2406.12863. 2024. Available from: https://arxiv.org/abs/2406.12863

[3] Goldston DA, Montgomery HL. Pair correlation of zeros and primes in short intervals. In: Adolphson AC et al., editors, Progress in MathematicsAnalytic Number Theory and Diophantine Problems. Vol. 70. Boston: Birkhäuser; 1987. DOI: 10.1007/978–1–4612–4816–310

[4] Petrzela J. Chaos in analog electronic circuits: Comprehensive review, solved problems, open topics and small example. Mathematics. 2022;**10**:4108. DOI: 10.3390/math10214108

[5] Deep AK, Raja GL, Meena GD. Robust moments-matching load frequency control strategy for cyber–physical power system amid communication time delay. Applied Energy. 2025;**382**:125251. DOI: 10.1016/j.apenergy.2024.125251

[6] Ansari ZA, Raja GL. Enhanced cascaded frequency controller optimized by flow direction algorithm for seaport hybrid microgrid powered by renewable energies. Applied Energy. 2024;**374**: 123996. DOI: 10.1016/j.apenergy.2024.123996

[7] Aryan P, Raja GL, Vilanova R. Equilibrium optimiser tuned frequency-shifted internal model control proportional-derivative decoupled dual-loop design for industrial plants followed by experimental validation. International Journal of Systems Science. 2024;**55**(14): 2874-2896. DOI: 10.1080/00207721.2024.2363544

[8] Kumari N, Aryan P, Raja GL, et al. Dual degree branched type-2 fuzzy controller optimized with a hybrid algorithm for frequency regulation in a triple-area power system integrated with renewable sources. Protection and Control of Modern Power Systems. 2023; **8**:48. DOI: 10.1186/s41601-023-00317-7

[9] Hu C, Hou Y, Bai J, Zou H. A new design of double predictive proportional integral control strategy for first-order plus dead-time industrial processes. The Canadian Journal of Chemical Engineering. 2024;**103**(3):1349-1362. DOI: 10.1002/cjce.25442

[10] Mehta U, Aryan P, Raja GL. Tri-parametric fractional-order controller Design for Integrating Systems with Time Delay. IEEE Transactions on Circuits and Systems II: Express Briefs. 2023;**70**(11):4166-4170. DOI: 10.1109/TCSII.2023.3269819

[11] Mukherjee D, Raja GL, Kundu P, Ghosh A. Design of optimal fractional-order Lyapunov-based model reference adaptive control scheme for CSTR. IFAC-PapersOnLine. 2022;**55**(1):436-441. DOI: 10.1016/j.ifacol.2022.04.072

[12] Aryan P, Raja GL. Restructured LFC scheme with renewables and EV penetration using novel QOEA optimized parallel fuzzy I-PID controller. IFAC-PapersOnLine. 2022;**55**(1):460-466. DOI: 10.1016/j.ifacol.2022.04.076

[13] Aryan P, Raja GL, Muduli UR. Optimal RIMC-based DLC strategy for

time-delayed higher-order integrating and unstable processes with a case study on inverted pendulum. In: 2023 IEEE 3rd International Conference on Smart Technologies for Power, Energy and Control (STPEC). Bhubaneswar, India: IEEE (Institute of Electrical and Electronics Engineers); 2023. pp. 1-6. DOI: 10.1109/STPEC59253.2023.10430588

[14] Aryan P, Raja GL, Vilanova R, Meneses M. Optimal internal model control-based decoupled dual-loop control method for boiler steam drum. In: 2023 IEEE 28th International Conference on Emerging Technologies and Factory Automation (ETFA). Sinaia, Romania: IEEE (Institute of Electrical and Electronics Engineers); 2023. pp. 1-4. DOI: 10.1109/ETFA54631.2023.10275712

[15] Kapitaniak T. Chaos for Engineers: Theory, Applications, and Control. Springer Berlin Heidelberg; 2000. DOI: 10.1007/978−3−642−57143−5

[16] Vivaldi F. Algebraic number theory and Hamiltonian chaos. In: Luck JM et al., editors. Springer Proceedings in PhysicsNumber Theory and Physics. Vol. 47. Berlin, Germany: Springer; 1990. DOI: 10.1007/978−3−642−75405−031

[17] Xing JT. Energy flow theory of nonlinear dynamical systems with applications. In: Emergence, Complexity and Computation. Berlin, Germany: Springer; 2015. DOI: 10.1007/978−3−319−17741−0

[18] Champneys A, Tsaneva-Atanasova K. Dynamical systems theory, bifurcation analysis. In: Dubitzky W et al., editors. Encyclopedia of Systems Biology. New York, NY: Springer; 2013. DOI: 10.1007/978−1−4419−9863−7270

[19] Ghys É. The Lorenz attractor, a paradigm for chaos. In: Duplantier B et al., editors. Progress in Mathematical PhysicsChaos. Vol. 66. Birkhäuser. Berlin, Germany: Springer; 2013. DOI: 10.1007/978−3−0348−0697−81

[20] Hidary JD. Table of quantum operators and Core circuits. In: Quantum Computing: An Applied Approach. Berlin, Germany: Springer; 2021. DOI: 10.1007/978−3−030−83274−215

[21] Lee H, Kim S. Entropy and predictability in dynamical systems with unimodal distributions. Chaos, Solitons, and Fractals. 2011;**18**(5):1019-1031

[22] Sorensen DC. Implicitly restarted Arnoldi/Lanczos methods for large scale eigenvalue calculations. In: Keyes DE et al., editors. Parallel Numerical Algorithms. Vol. 4. Springer, Dordrecht: ICASE/LaRC Interdisciplinary Series in Science and Engineering; 1997. DOI: 10.1007/978−94−011–5412–35

[23] Wolf M. Will a physicist prove the Riemann Hypothesis? 2014. Available from: https://arxiv.org/abs/1410.1214

[24] Tse CK. Design-oriented bifurcation analysis of power electronics systems. In: Valentin IN et al., editors. Understanding Complex SystemsApplications of Nonlinear Dynamics. Berlin, Germany: Springer; 2009. DOI: 10.1007/978−3−540−85632−014

[25] Dabrowski A, Balcerzak M, Pikunov D. Improving efficiency of the largest Lyapunov exponent's estimation by its determination from the vector field properties. Nonlinear Dynamics. 2020; **102**:1869-1880. DOI: 10.1007/s11071−020−05994−9

[26] Sharkovsky AN, Maistrenko YL, Romanenko EY. Periodic trajectories. In: